Curious Behavior

curious behavior

YAWNING, LAUGHING, HICCUPPING, and BEYOND

Robert R. Provine

THE BELKNAP PRESS *of* HARVARD UNIVERSITY PRESS

Cambridge, Massachusetts

London, England

Printed in the United States of America

First Harvard University Press paperback edition, 2014

Library of Congress Cataloging-in-Publication Data
Provine, Robert R.
Curious behavior : yawning, laughing, hiccupping, and
beyond / Robert R. Provine.
p. cm.
Includes bibliographical references.
ISBN 978-0-674-04851-5 (cloth: alk. paper)
ISBN 978-0-674-28413-5 (pbk.)
1. Human behavior. 2. Human biology. 3. Neuropsychology.
4. Evolutionary psychology. I. Title.
BF199.P765 2012
152.3′2—dc23 2012007754

To Helen

contents

Curious Behavior

introduction

We humans are talkative, sociable, bipedal, tool-using mammals that Shakespeare found noble in reason and infinite in faculty. The Bible tells us that we are made in God's own image. We have walked on the Moon, invented the computer, and analyzed our own evolution. But humans are also farting, belching, yawning, hiccupping, coughing, laughing, crying, sneezing, vomiting, itchy, scratching, ticklish herd animals. These beastly, instinctive acts help to define us as a species, but they are neglected by scientists who overlook the familiar. Biologists usually focus on general processes in physiology or genetics, not specific, unheralded acts. Social scientists attend to environmental influences on behavior, ignoring the instincts that are the bedrock of our being. Physicians treat symptoms such as sneezes or coughs but seldom study them. Some priggish dictionaries spare us farts, even in printed form.

This book redresses historic debts, putting selected instincts front and center in an analysis and celebration of undervalued, informative, and sometimes disreputable human behavior. It adds new topics to the scientific agenda, views the familiar in new ways, and shows how such curious acts can be tools to examine broader and deeper issues, from the origin of movement to a new approach to the unconscious. However, in championing the new or neglected, our enthusiasm should be tempered by Carl Sagan's wise counsel: "They laughed at Columbus, they laughed at Fulton, they laughed at the Wright brothers. But they also laughed at Bozo the Clown" (*Broca's Brain*, 75). Although some of the behaviors considered here are obscure, titillating, or bizarre, all have earned the right to scientific scrutiny. The performance of an act by people of all sexes, races, and cultures is reason enough for its study—it's a part of our biological heritage and has a low Bozo factor. But there are motives beyond telling our species' untold tales.

Scientific advances often come from analyses of the elemental, whether molecule, bacterium, or fruit fly, with the expectation that their study will illuminate more complex, less manageable systems. This *simpler system* approach continues here, but the level of reduction is boosted up a few notches from molecules to elemental human behavior. We learn, for example, that a vibrating membrane is involved in sound-making, whether aria or fart, and why the vocal, not the alimentary, tract was selected as the more flexible instrument for auditory signaling. There are very good reasons why we don't engage in buttspeak.

Along the way, we learn that the simple act of tickling provides the neurological mechanism for the computation of self and other—the central and unappreciated event in the emergence of personality and social behavior, and the starting point for programming personhood into robots. Who would anticipate that tickling provides a bridge between the often estranged social, neurological, and computational sciences? Or that the itch/scratch system joins tickling in having its own social dimension? Tickling also provides the neurological mechanism of physical play, whose program binds us together in the give-and-take of rough-and-tumble and sex play.

We are part of a human herd whose behavior is often the involuntary playing out of an ancient neurological script that is so familiar that it goes unnoticed. Consider what is really happening when your body is hijacked by an observed yawn or you spontaneously join others in a communal chorus of ha-ha-ha. You don't decide to yawn or laugh contagiously—it just happens. We discover that a simple neurological mechanism that automatically replicates observed behavior is the basis for contagious yawning, laughing, crying, scratching, and coughing, and perhaps sociality and empathy. American president Bill Clinton's celebrated ability to "feel your pain" may be related to his joining friends in a yawn or laugh. A malfunction of this mechanism of contagious behavior may be responsible for the social deficits of autism and schizophrenia.

Laughter is revealed to be the sound of physical play, with the labored pant-pant of our cavorting primate ancestors evolving into the human ha-ha, the best example of the origin of a vocalization. Laughter also shows why we can talk and chimpanzees can't, and why the breath control conferred by bipedality is essential for this transformation.

Emotional tears are a breakthrough in human social evolution, adding range and nuance to the face as an instrument of emotional expression. Will people recognize your sadness without tears? The evolution of the white of the eye (the sclera) is another transformative social event. We learn why the uniquely white human sclera is necessary for the display of red eyes, a signal of sadness and poor health, and why eyedrops that "get the red out" are really beauty aids.

Strangest of all, we begin life as an embryo, a maternal parasite whose seizure-driven behavior defies psychological convention and forces a rethinking of the causes and functions of prenatal behavior. No textbook of child psychology provides a useful account of prenatal behavior, describes its origin in spontaneous spinal cord discharges (not the brain), or catalogues its exciting roles in regulating neuron numbers, sculpting joints, stretching the skin and placenta to fit, and fine-tuning neural circuits.

Mother Nature Does Not Reinvent the Wheel

Organisms are collections of biological clichés. Therefore, we can use one organism to understand another, the rationale of *comparative analyses.* Whether the process is development, genetics, or physiology, there are a limited number of ways to get from point A to point B. These successful processes evolved gradually by means of the relentless engine of natural selection, through which individuals best adapted to their surrounds enjoy increased reproductive success, passing their traits—adaptive, maladaptive, and neutral—to descendants who gradually constitute a greater proportion of the population. Once evolved, these successful solutions are used over and over. One species can contribute to our understanding of others because they share a similar biology and heritage. Science would be unwieldy if each species had its own, unique genetics, development, and physiology. In a hypothetical world of such biological uniqueness, there would be few organisms, none complex, and we would not be around trying to figure these things out.

This book explores a variant of the comparative approach, one that contrasts behaviors instead of organisms. This tactic yielded some surprises. The yawn resembles a slow sneeze (or the sneeze a fast yawn). The yawn starts with a long inspiration and shorter exhalation, whereas the sneeze starts with a faster inhalation and explosive, noisy, exhalation. But the story gets better. The facial expression builds during the initial stage of both yawn and sneeze, relaxing during the climactic exhalation. If the yawning/sneezing face seems vaguely familiar in another context, you have seen it as part of an orgasm. The underlying principle is significant—that multiple behaviors tap a similar neuromuscular substrate, and understanding one helps us understand the others.

Modern humans are treated as a mixture of the ancient and the new, hunter-gatherers trying to make sense of their actions and cosmic predicament using a quirky brain that may not quite be up to the task. We are reminded that evolution does not proceed like a neurological urban renewal project, where the new emerges de novo on the rubble of the old. Instead, the new is tweaked, refined, and jury-rigged from odds and ends in our neurological basement and is erected on the old foundation, which stays in place. The ancient neurological circuitry remains, sometimes useful, other times useless, quiescent, unruly, or out of control.

The present sample of quirky behaviors is a rich hunting ground for such neurological and behavioral fossils, whether surviving intact or as vestiges or building blocks of modern behavior. Pursuit of such fossils adds an air of exoticism and adventure to the mundane, and does not require a field trip to a remote and desolate archeological dig.

The evolutionary events that shaped us are not part of a distant and unknowable past because we carry the imprint of our biological heritage. We do not pass through generations shedding our biological endowment as we move on. Whatever we have been, in some way we are still, and remnants of the past leave subtle but undeniable traces. As William Faulkner reminded us in *Requiem for a Nun*: "The past is never dead. In fact, it's not even past."

Ethology and the New Unconscious

The scientific approach that most influenced this research is ethology, the evolutionarily based, biological study of behavior. Ethology emerged from the European tradition of field studies and natural history and was pioneered by Konrad Lorenz, Nikolaas Tinbergen, Karl von Frisch, and their nineteenth-century forebear, Charles Darwin. Ethology has a naturalistic focus, typically observing behavior in its natural setting, and is more concerned with the adaptive (evolutionary) significance of behavior than is typical of American psychology, with its greater focus on learning, environmental influences, and carefully controlled laboratory experiments.

Ethologists study instincts such as species-specific calls and mating behaviors, or the present, nontraditional fare of human sneezes, yawns, coughs, and the like. For those wary of just-so stories of a fanciful evolutionary psychology, be assured that the present treatment is anchored in concrete, observable acts of the here and now. Not long ago, instinct was the forbidden *i*-word that was not mentioned in polite scientific company, especially among many social scientists. As noted by Steven Pinker in *The Blank Slate: The Modern Denial of Human Nature*, many behaviorists and other social scientists still reject the existence of instincts and other biological foundations of human nature. For a variety of philosophical, political, and scientific reasons, the acceptance of instincts was considered a step down a slippery slope that would presumably lead to biological determinism, social Darwinism, sexism, racism, and unnamed isms too horrible to contemplate. Ironically, the analysis of instinctive behaviors can be unifying, by focusing on human universals instead of differences between individuals and cultures.

The unlikely trio of Sigmund Freud, Carl Jung, and B. F. Skinner influence the present work in subtle ways, but their lessons may not be the ones that they intended to teach. Freud's influence lurks in his stress on unconscious (involuntary) processes in human behavior, although the present account is decidedly not psychoanalytic. Carl Jung's influence is found in the search for ancient, instinctive behavior, but my human "archetypes" are certainly not Jungian. The behavioral approach of B. F. Skinner, which focuses on what people do instead of what they claim as their motives, proves useful for studying unconsciously controlled human behavior, a Freudian theme. Marx was also an influence—Groucho, not Karl.

People, we will find, are inaccurate reporters of their actions, especially the spontaneous and involuntary sort considered in this book. With apologies to Freud, Jung, and associates, the semantic quagmire of defining "voluntary" or what is "consciously controlled" is finessed by defining them in terms of reaction times. You are presumed to have more voluntary control of behaviors that, upon verbal command, can be performed with the shortest reaction times. The piano-like display of the Behavioral Keyboard in the appendix summarizes the approach and provides reaction times of ten common behaviors considered in various chapters, with its rightmost key for the rapid eye blink and the leftmost key for the tardy cry. Keys toward the far left, including crying, hiccupping, sneezing, and yawning, are extremely sluggish, if they can be played at all. On this clunky instrument, even Beethoven couldn't pound out blink-blink-blink-sneeze.

The full cast of characters who contribute to this exploration extends far beyond Freud, Jung and Skinner, dates from antiquity, and grows chapter by chapter. Human universals have been the topic of study and speculation since Plato and Aristotle, but undocumented research on these topics is certainly much older, extending to the beginning of human self-reflection. Understanding the behavior of self and others is, after all, a matter of life and death.

In Praise of Small Science

Big Science is exemplified by the Large Hadron Collider, the giant particle accelerator in Europe, an instrument so large that it ranks among humankind's most formidable artifacts, so expensive that it must be funded by many nations, and so complex that it must be run by a giant institute with thousands of employees. If you want to do cutting-edge particle physics and are not affiliated with the project, your career may be on the slow track. Such physically imposing instruments look expensive and important. They are. This book is about a very different kind of science.

The Small Science of this book is "small" because it does not require fancy equipment and a big budget, not because it's trivial. Some of it can be done with pen and paper and involves nothing more than taking notes at social gatherings or at the local mall. A cadre of research assistants is unnecessary, and useful observations can be made using yourself as subject. If you insist on more equipment, buy a stopwatch. This research can be self-funded, allowing you to happily avoid grant writing and the dispiriting, multiyear competition for ever-shrinking funds. Good thing! Such research lacks scientific grandeur and is unlikely to excite funding lust in grants administrators and politicians.

The research techniques of Small Science are often simple, eliminating another barrier for participation: the months or years needed to master new techniques. When possible, anything worth doing is worth doing quickly. Short start-up times and the expectation of timely project completion increase the fun. Even monks toiling in the scientific vineyard can benefit from an occasional shot of adrenaline and timely reward. I know that I do.

The low barrier for entry keeps practitioners of Small Science on their toes. This is science of the most democratic kind. It's not quantum electrodynamics. Everyone begins as something of an expert, having lifelong experience with yawns, sneezes, and the like, and will be watching with an informed and critical eye. When presenting results, you had better get things right because there will be a posse of enthusiasts in hot pursuit, anxious to advance, confirm, or challenge.

Even the scientific literature is relatively accessible. Unlike arcane topics in mathematics or theoretical physics, you can detect insight, confusion, puffery, or flapdoodle. Like most readers, I appreciate straight, jargon-free talk about everyday things, and abhor florid "neurologizing" and "biologizing," the dressing up of behavioral accounts in the trappings of neurology and biology to provide an illusion of depth and substance. Big words and complicated concepts must earn their keep. Simplicity makes people feel smart and competent, and they behave accordingly, becoming partners in the adventure. Unnecessary complication has the opposite effect.

Small Science, if successful, may not remain small. Once the value of the long-neglected is appreciated, be ready for the stampede of Big Science headed your way, armed with big budgets, fMRIs, and the certitude that they were there at the conception. If this should happen, you can retire from the fray and look back wistfully at the golden age when things were simple and the frontiers near at hand, join the Big Science crowd, or, like a prospector of the Old West, move on, looking for the next mother lode. I identify with the old prospector who is motivated as much by the search as by the discovery, mapping the terrain for those who wish to follow, whether or not it leads to the Next Big Thing.

Sidewalk Neuroscience

This book is full of *sidewalk neuroscience,* a scientific approach to everyday behavior based on simple observations and demonstrations that readers, even advanced gradeschoolers, can use to confirm, challenge, or extend the reported findings. Potential science fair projects as well as PhD dissertations are found in these pages. Clear description, the core of this approach, never goes out of style, and its pursuit has a high bird-in-hand factor, meaning that the least desirable outcome is still pretty good. Some explorations are urban safaris to suburban shopping malls, city sidewalks, or dinner parties instead of the tropical rain forest—field trips without the expense and biting insects. The behaviors considered in this book are exemplars of works in progress, not an exhaustive list. Everyday life is teeming with the important and unexpected, if you know where to look and how to see.

I've done plenty of bench science of the more traditional kind, from tissue culture to neurophysiology. While I fancy the gadgetry and techniques of my electrophysiology lab, times and people change. I now enjoy getting out of my windowless lab and communing with fellow human beings using techniques that my students can easily master, and indulging my taste for the unconventional. This austere, low-tech approach is not for everyone. I recall the polite skepticism of a few of my early research students about the nontraditional topic of yawning, who wondered, "If yawning is such a big deal, why aren't more scientists working on it?" or "If the project is so important, why am I a member of the research team?" I urged them to resist Groucho Marx's declaration—"I would never join a club that would have someone like me as a member." If it's fun and has promise, forge on. Let posterity weigh its ultimate worth.

This account describes hard-won insight, false leads, misplaced enthusiasm, disappointment, time-consuming diversions, and the occasional serendipitous discovery. Even apparent failure can be success in disguise, leading to a better direction. Piece by piece we collect parts of a scientific puzzle that, once assembled, will help us see the commonplace in new ways and reveal a perspective on human nature that was always hidden in plain sight.

1

yawning

We steer our body through life's straits and shoals, walking, working, talking, speeding up and slowing down, avoiding obstacles. We are captains of our ship, alert, confident, and rational. That is the illusion. But what if we are deceived by our brain's subtle whispers, its effort, as in dreams, to weave a coherent, sometimes faulty narrative from irrational events? Are we instead unthinking herd animals, driven by subconscious instincts, acting out our species' ancient biological script? Pursuit of this theme requires rethinking the human condition and turning history on its head, immodest goals for a chapter about yawning. We will settle instead for revealing chinks in our neurologically generated, virtual edifice of daily life. Turning history on its head must wait for another day.

Imagine the face of a yawning person, with gaping jaws, squinting eyes, and a long inhalation followed by a shorter outward breath. Ahh. This visual stimulus hijacks your body and induces you to replicate the observed behavior. As many readers have noticed, the contagiousness of yawns is so potent that simply discussing yawns triggers yawns.[1] Contagious yawns occur automatically, without any desire to imitate a yawner. When you see someone yawn, do you think, "I want to yawn just like that person"? That this remarkable phenomenon has not received more scientific scrutiny is testament to our undervaluing the familiar. But this situation is starting to change. The implications of behavioral contagion are so broad and deep that they transcend disciplinary boundaries and include anyone interested in the roots of human social behavior, including those who couldn't care less about yawning.[2]

The Doomsday Yawn

The Doomsday Yawn is the supremely powerful mother of all yawns, whose contagion is so great that it renders us helpless under its spell, slaves to bouts of irresistible yawns. My motive for creating a Doomsday Yawn was scientific, not megalomaniacal. The first step in designing a super-yawn is to understand the parts of a normal yawn that are the vector for ordinary contagion, then maximize their effect. The idea is borrowed from a Monty Python skit about a Doomsday Joke that is so potent that it inflicts a lethal fit of laughter upon adversaries on the battlefield—the joke's punch line is not added to the setup until immediately before transmission to avoid the accidental exposure of friendly forces to its devastating effects. Although a weaponized Doomsday Yawn is an equally meager tool for warfare, the idea of an artificial, super-powerful (supranormal) stimulus is well established in animal behavior research and guided our work. The ability to create a potent yawn stimulus is valuable because it requires understanding of the stimulus vector.

Before launching a Manhattan Project for a Doomsday Yawn, it's important to confirm the folklore that yawns really are contagious. This was done by exposing observers to a five-minute series of thirty videotaped repetitions of a male adult yawning; one six-second videotaped yawn was presented every ten seconds.[1] Yawns are clearly contagious. Observers were more than twice as likely to yawn while observing the yawns (55 percent) than to yawn while viewing a comparable series of smiles (21 percent), the control condition. These results are highly reliable; other laboratories confirmed that contagious yawns occurred in about half of observers.

Yawning does not involve a simple reflex like the knee jerk, in which the response occurs almost immediately after the stimulus (a tap to the patellar tendon) and is proportionate in strength to the stimulus. Evoked, contagious yawns appeared at varying latencies throughout the five-minute test period, and when they occurred, they had *typical intensity*, meaning that they were regular in form and vigor.[1] Yawns are more complex in structure, have slower and more variable onset, and last longer than classical reflexes such as the knee jerk. In the language of classical ethology, yawning is a *stereotyped (fixed) action pattern* that is *released* by the *sign stimulus* of the observed yawn. (I prefer the term "stereotyped" to "fixed" to describe such motor patterns because it suggests a strong central tendency, not the indefensible straw-man position of invariance.) More will be said later about yawning as a motor act.

Nature conspires to maximize contagiousness. Our set of videotaped yawn stimuli proved equally potent whether viewed right side up, sideways, or upside down—the brain's yawn detector is not axially specific.[3] Furthermore, the subjects' yawn detector is dependent neither on color nor on movement, because the videos are equally potent when viewed in color or high-contrast black and white, or when the usually animate stimulus is presented as a still image of the yawner in mid-gape.

Next, the features of the yawning face necessary to prompt yawns were investigated,[3] a critical step in engineering the Doomsday Yawn. There were some big surprises. Most people incorrectly presume that the gaping mouth is the signature of the yawn, but animate, yawning faces edited to mask the mouth are just as effective in producing yawns as the intact face (Fig. 1.1). The finding of potent, mouthless faces was initially troubling. Had something gone wrong in this big, labor-intensive study with 360 participants? I was relieved to find complementary data showing that the mouth was not necessary; the disembodied yawning mouth is no more effective in evoking yawns than the control smile. Outside the context of the yawning face, the open mouth is an ambiguous stimulus—the mouth could equally well be yelling or singing. The most obvious trait of the yawning face is not a vector of its contagion. Our neurological yawn detector responds to the overall pattern of the yawning face, including the squinting eyes, and not a particular facial feature. Stretching and postural changes in the upper body (e.g., tilting of head, raising of shoulders) may provide other cues.

Miss Manners take note: this incidental contribution to etiquette research suggests that shielding your mouth is a polite but futile gesture that will not prevent the passing of your yawn to companions. It also suggests why it's difficult for artists to portray yawns in their paintings and cartoons—the cliché of the gaping mouth is insufficient to define a yawn, and supplemental context cues such as a hand shielding the mouth and outstretched arms are often used to reduce ambiguity. Sometimes cartoonists surrender to the challenge and write "YAWN" in a bubble over the character's head.

The yawn is so potent that even thinking about it triggers yawns. In fact, thinking about yawning was the yawn-induction procedure used in several of my studies, evoking yawns in up to 92 percent of subjects within thirty minutes.[1,4]

figure 1.1 *What facial features prompt contagious yawns? Participants who viewed videos of a series of normal yawns were about twice as likely to yawn as those who viewed a series of smiles. When the faces were edited to test which features were most potent in prompting yawns, the disembodied gaping mouth was not a very effective stimulus. Yawning faces whose mouths were masked were as effective as intact faces in producing yawns. (From Provine 2005)*

figure 1.2 *People reading about yawning yawn more often than those reading about hiccupping, another unconscious act. Hiccupping is not contagious. Yawning and hiccupping (Chapters 1 and 8) are subject to social inhibition, a fact that distinguishes them from many, but not all, other unconscious actions. (From Provine 2005)*

And as many readers have noticed by now, reading about yawning also triggers yawns. When put to the test, 28 percent of subjects who read an article about yawning for five minutes reported yawning during this period, versus 11 percent of a control group who read an article about hiccupping (Fig. 1.2). When the criterion was relaxed to include those who either reported yawning or being tempted to yawn, the difference between the yawn and hiccupping conditions grew to 76 percent and 24 percent, respectively, still about a threefold difference.

The discovery of the global range of yawn triggers reduced my enthusiasm for the Doomsday Yawn project; a career can be made of exploring the long list of possible yawn stimuli. No single stimulus dimension has the potential of being tweaked into a super-yawn, and the maximum short-term rate of yawn production is around 50 percent. I could not, as planned, produce the perfect, irresistible yawn by synthesizing a gaping mouth of just the right size and shape that opened and closed at just the right rate.[5] Disappointment was tempered by the revelation that we already possess a pretty good approximation of a Doomsday Yawn. Even the sigh-like sound of yawns can trigger yawns in visually normal[6] and blind individuals,[7] and neutral stimuli can acquire yawn-inducing properties through association. My reputation as a yawn sleuth has conferred a curious kind of charisma—I've become a yawn stimulus.

The Roots of Sociality

Yawns are propagated, being passed from one person to another in a behavioral chain reaction. This mindless connectedness involves social behavior of the most primal sort. The rippling of a behavior through a group is heritable, neurologically programmed social behavior of a sort neglected by social scientists who often explain behavior in terms of learning during the lifetime of the individual. Yet contagion is a factor in the development and evolution of sociality and perhaps empathy.[8]

Interest in sociobiological phenomena related to contagion has been prompted by the recent discovery of "mirror neurons" in the premotor cortex of monkey brains by Giacomo Rizzolatti and a team of neuroscientists at the University of Parma in Italy.[9] These neurons are active both when a monkey performs an act such as grasping a peanut and when the monkey observes the same act being performed by others. Noninvasive imaging (fMRI) has detected similar mirror activity in several regions of the human brain (premotor cortex, supplementary motor area, primary somatosensory cortex, and inferior parietal cortex).[9]

Neuroscientists posit a role for mirror neurons in many behaviors, including imitation, intuition, empathy, language, and theory of mind (our appreciation that other people have minds like our own), and suggest that "broken mirrors" may be responsible for the social deficits of autism.[10] V. S. Ramachandran of the University of California, San Diego, could hardly be more enthusiastic in his essay "Mirror Neurons and Imitation Learning as the Driving Force Behind the Great Leap Forward in Human Evolution," on Edge.org: "I predict that mirror neurons will do for psychology what DNA did for biology: they will provide a unifying framework and help explain a host of mental abilities that hitherto remained mysterious and inaccessible to experiment." No small order, that!

As a research topic, contagious behavior has similar promise but offers a significant advantage over mirror neurons: an easily measured output such as yawning. At present, mirror neurons are like disembodied computers not hooked up to printers—full of potential, but short on demonstrated function. Curiously, mirror neuron researchers have almost nothing to say about the relation between mirroring and contagious behavior (is it insufficiently cognitive?), so we must speculate whether contagion involves a mirror process or something completely different. Preliminary fMRI data indicate that the brain areas that respond uniquely to observed yawns are the same association areas linked directly or indirectly to theory of mind and self-processing, but the details are still being worked out.[11]

Comparative, developmental, and pathological analyses offer other approaches to contagious yawning and its mechanism. From them, we can learn when contagion evolved, when it develops, and the effects of malfunction.

Little is known about contagiousness in creatures beyond human adults, but this situation is starting to change. James Anderson and his colleagues at the University of Stirling in

Scotland report contagious yawning in chimpanzees,[12] primates that show rudimentary empathy and self-awareness as reflected in mirror self-recognition tasks.[13] Contagion, if present, may be weaker in monkeys (macaques and baboons)[14] and other animals that are deficient in these traits of social cognition. The ability of dogs, a highly attentive companion animal, to catch yawns from humans and other dogs is the topic of conflicting reports.[15] It's obvious that contagion operates from animal to human, given that any stimulus associated with yawning can trigger its contagion.

Contagious yawning appears relatively late in human development. Although fetuses yawn spontaneously in the womb and yawns are obvious in newborns, James Anderson and Pauline Meno did not detect contagious yawning in response to videos of yawns in children until after five years of age. Using live instead of video yawns, Molly Helt and colleagues at the University of Connecticut observed contagious yawns as early as two years, with significant increases at four years.[16] This relatively late onset, combined with its rarity in nonhuman species, suggests that contagious yawning has an evolutionary origin separate from and more recent than the very ancient and ubiquitous act of spontaneous yawning.

Malfunctions of the contagious yawning mechanism may be responsible for symptoms of psychopathology and neuropathology. Atsushi Senju of Birkbeck College at the University of London and colleagues showed that children with autism spectrum disorder are deficient in contagious yawning, although their ability to perform spontaneous yawning is spared.[17] The diminished contagion may be associated with their impaired ability to empathize and form normal emotional ties with other people. Contagious yawning can be increased if autistic individuals are instructed to observe the eyes, an important cue for contagious yawning.[18]

Contagious yawning may also be useful in understanding aspects of schizophrenia, a severe psychiatric disorder associated with the inability to infer the mental states of others. Steven Platek, then at the University of Albany, and colleagues examined contagious yawning in people who are not clinically ill but are schizotypal—that is, who are deficient in their ability to empathize with what others want, know, or intend to do, and who have certain other problems in thought and behavior.[19] Schizotypal individuals have reduced susceptibility to contagious yawning. Helene Haker and Wulf Rossler at the University of Zurich established that schizophrenic individuals showed lower than normal levels of both contagious yawning and laughter.[20] Provocatively, the late Heinz Lehmann claimed that increases in yawning (contagious yawning was not specifically examined) predicted recovery in schizophrenia.[21]

A challenge of such contagion research, whether using young or old humans, normal or pathological humans, or other animals, is in maintaining the attention of distractible subjects. Without careful controls, the apparent lack of infectious yawning may be only an artifact of differential inattention, not evidence of a deficit or difference in cognitive or social function. Positive results are, therefore, more persuasive than negative ones.

Unconscious Control

You can yawn spontaneously or contagiously, but you cannot yawn on verbal command. Your inability to yawn voluntarily is evidence of its unconscious control. We used reaction time to objectively determine and contrast our conscious control of yawning and other behaviors. You are presumed to have greater conscious control over behaviors with short reaction times than those with long reaction times.

With stopwatch in hand, lab members Kurt Krosnowski, Nicole Whyms, Megan Hosey, and Cliff Workman measured the time required to yawn on command. Each of our 103 participants was instructed to yawn after being told "Now," when the experimenter started the watch. Experimenters recorded either the latency of yawn initiation or a maximum of ten seconds if the act could not be performed within a ten-second time limit. Yawns had a sluggish average reaction time of 5.7 seconds and were hard to produce; 42 percent of subjects did not even attempt them, receiving the maximum score of ten seconds. The large latency and low rate of successful yawns (most were probably fakes) indicate a low level of conscious control. The reaction times of yawns and other acts discussed in subsequent chapters are presented graphically via the Behavioral Keyboard in the appendix. It is a piano-like display that ranks response latencies (and conscious control), from the speedy, highly controllable eye blink (0.5-second reaction time), on the far right key, to the lethargic, uncontrollable cry (9.8-second reaction time), on the far left key.

Outside the lab we got further evidence of the fragility of the contagious response, its unconscious control, and the significance of social context. Intense self-awareness, as when you are being observed or even suspect that you may be observed, inhibits yawning. This was the basis of my rationale for the use of self-report of yawning via button pressing instead of visual observation or video recording.[1] Because my yawning studies attracted the attention of the popular media (they have a voracious appetite for such stories), I had the experience of seeing the inhibition in action.

A television newsmagazine crew turned up one day to tape a segment for their show. Against my advice, the show's producer set out to re-create my experiment, in which one-half of a large lecture class read an article about yawning while the other half read a control passage about hiccupping. Normally, the effect of the yawning article is robust and has been used as a demonstration of contagion in classes at other universities. It's amusing to see one-half of a large lecture class yawning while the other half is not.

As I predicted, the class demonstration did not survive the up-close-and-personal scrutiny by a national network television crew with their intrusive cameras, and only a tiny fraction of the usual amount of yawning was observed. The crew performed an unintentional but informative variant of my original research that demonstrated the powerful effect of social inhibition on yawning.

Even highly motivated, prolific yawners who learned how to yawn by thinking about yawns, who practiced on their own, and who volunteered to be taped individually for the national television show stopped yawning when placed before the camera. "Hey, kid, want millions of people to see you yawn?" My confident volunteers seemed amazed that something so easy when waiting in the hallway a few minutes before became so difficult during the moment of truth. The social inhibition of yawning occurred spontaneously under scrutiny and was not the voluntary effort of the yawner to suppress a rude or inappropriate act. The socially significant act of yawning can be either produced or inhibited by unconscious processes.

The sensitivity to scrutiny highlights a general neurobehavioral principle about yawning and behaviors covered in later chapters. The phylogenetically old does not meld seamlessly with the new. When the unconscious and the conscious compete for the brain's channel of expression, as when yawning under social scrutiny, the more modern, conscious mechanism suppresses its older, unconscious rival.

The Act of Yawning

The verb "to yawn" is derived from the Old English *ganien* or *ginian*, meaning "to gape" or "to open wide" (chasms really *do* yawn). When yawning, you join vertebrates everywhere in one of the animal kingdom's most ancient rites. Mammals and most other animals with backbones yawn; turtles, crocodiles, snakes, birds, and even fish do it.[22] Yawning is present by the end of the first trimester of prenatal human development and is obvious in newborns.[23] The early development of yawning is evidence of its evolutionary antiquity.

During a yawn, you stretch your jaws open in a wide gape, take a deep inward breath followed by a shorter exhalation, and end by closing your jaws.[1] But yawning has significant features in addition to gaping jaws that are easy to observe and analyze. Yawns are easy to collect by tapping the just-described contagion response. Just ask people to think about yawning and they will produce genuine yawns.

In many of my studies, I asked participants sitting in an isolation chamber to think about yawning, and to push a button at the start of a yawn and keep it depressed until they finished exhaling at the end of a yawn. Self-report was used because, as noted above, yawning is inhibited in subjects who think they are being observed by experimenter or camera.

Here are some of the things that I learned. The yawn is highly stereotyped but not invariant in duration and form.[1] It is an excellent example of the instinctive stereotyped (fixed) action pattern of classical ethology. As noted previously, it is *not* a reflex, a simple, short-duration, proportional response to a simple stimulus. Once started, a yawn progresses with the inevitability of a sneeze. The yawn runs its course in about six seconds on average, but its duration can range from about three and a half seconds to much longer than the average. There are no half yawns, an example of the "typical intensity" of stereotyped action patterns; this is also a reason why you cannot stifle a yawn. When you feel a yawn coming on, you might as well go along for the ride. Yawns often come in bouts, with a highly variable inter-yawn interval (onset to onset) of around sixty-eight seconds. There is no relationship between yawn frequency and duration; producers of short or long yawns do not compensate by yawning more or less often. If you are now yawning, you can use a stopwatch to measure the duration of your yawns and the intervals between them.

I offer four informative yawn variants that test hypotheses about the form and function of yawning. You can test yourself and draw your own conclusions about yawning and its underlying mechanism. However, not everyone, including my long-suffering wife, shares my enthusiasm for such self-experimentation. And even enthusiasts may want to conduct the experiments in private. Let's begin.

The closed- nose yawn.[1] When you feel yourself start to yawn, pinch your nose closed (Fig. 1.3). Most participants report the ability to perform perfectly normal closed-nose yawns. This indicates that the inhalation at the onset of a yawn and the exhalation at its termination need not involve the nostrils—the mouth provides a sufficient airway. No surprise here, because you can breathe equally well through mouth or nose.

Now let's test some more interesting propositions about the role of the mouth and jaw.

The clenched-teeth yawn.[1] When you feel yourself begin to yawn, clench your teeth but permit yourself to inhale normally through your open lips and clenched teeth (Fig. 1.3). This diabolical variant gives many people the sensation of being stuck in mid-yawn, or at least of being unable to experience the relief of completing a yawn. We are now getting down to the nitty-gritty of yawn science. This experiment shows that the gaping of the jaws is an essential component of the

figure 1.3 *The physiological properties of a yawn can be determined with self-experimentation. A normal yawn* (upper left) *involves gaping jaws, a deep breath, and a shorter exhalation. If you pinch your nose shut when you begin to yawn* (upper right), *you will find that you yawn quite normally; the nostrils are not necessary for the deep inhalation. A clenched-teeth yawn* (lower left), *by contrast, is nearly impossible, revealing the yawn to be a complex motor pattern requiring gaping jaws, although inhalation remains possible through the teeth. If you try a nose yawn* (lower center)*—inhaling only through the nose—you will probably find it impossible. For most people, inhalation through the mouth is an essential component of the yawn. The open-eye yawn* (lower right), *accomplished by holding your eyes open with your fingers, is also difficult, indicating that feedback from the normally squinting eyes is essential to the performance of the motor pattern of yawning, although it has no association with the airway. (Modified from Provine 2005)*

complex motor program of the yawn; unless accomplished, the program will not run to completion. The yawn is also shown to be more than a deep breath because, unlike normal breathing, the inhalation and exhalation of the yawn cannot be performed equally well through the clenched teeth or the nose.

The sealed-lips "nose yawn."[24] This variant tests the adequacy of the nasal airway to sustain a yawn (Fig. 1.3). (The closed-nose yawn already showed the nasal airway to be unnecessary for yawns.) Unlike normal breathing, which can be performed equally well through mouth or nose, yawning is difficult, if not impossible, for most people to perform via nasal inhalation alone. As with the clenched-teeth yawn, the nose yawn provides the unfulfilling sensation of being stuck in mid-yawn. Inhalation through the mouth is an essential component of the motor pattern of yawning. Exhalation, on the other hand, can be accomplished equally well through nose or mouth.

The eyes-open yawn. This new yawn variant is reported here for the first time. When you feel yourself begin to yawn, prop open your eyes with your fingers and keep them open during the course of the yawn (Fig. 1.3). Although this latest variant is still under investigation, most participants report difficulty yawning with eyes open, noting that their yawns were either blocked or stalled. If you are questioning the relevance of the eyes to the act of yawning, recall that the eyes squint or close during yawns, and that observing squinting eyes contributes to the contagiousness of yawns.[3] (Squinting eyes and similar facial behavior are also present during the non-contagious acts of sneezing and orgasm; see Chapter 7.)

So far we have demonstrated that inhalation through the oral airway and the gaping of the jaws are essential for normal yawns, and that the motor program for yawning will not run to completion unless these parts of the program have been accomplished. But yawning is a powerful, generalized movement that involves much more than airway maneuvers, jaw gaping, and eye squinting. When yawning, you also stretch your facial muscles, tilt your head back, squint or close your eyes, tear, salivate, open the Eustachian tubes of your middle ear, and perform many cardiovascular, neuromuscular, and respiratory acts. So many muscles are active during yawning that it may be simpler to list which muscles are not involved.

~

Yawning probably shares components with other behavior, all being assembled from a neurological parts bin of ancient motor programs. For example, is the yawn a kind of "slow sneeze," or is the sneeze a "fast yawn"? Both share common respiratory and motor features, including jaw gaping, eye closing, head tilting, and tearing. A still photograph (but not a video) of one can be confused for the other. For me, both cataclysmic sneeze and gentler yawn start with a strange sensation in the upper back part of the throat.

Yawn- and sneeze-like facial expressions are also present during orgasm and all three end in climax, suggesting that the acts share a common neurobehavioral heritage.[24] For examples of facial expressions during an orgasm, visit beautiful agony.com. This imaginative website provides video images of hundreds of males and females masturbating to climax. Squeamish readers will be relieved to know that the images are strictly from the shoulders up, yielding images that are hardly more risqué than Meg Ryan's memorable fake orgasm scene ("I'll have what she's having") in the film *When Harry Met Sally*. The link between yawning and orgasm is not as farfetched as it sounds on first hearing, and it goes beyond superficial behavioral resemblance. For his exhaustive and unusually creative dissertation in art history at the Free University of Amsterdam, Wolter Seuntjens tracked down the surprisingly extensive and scattered connections between yawning and sex (but not sneezing) in science, the arts, and popular culture.[25]

Among most mammals, males are the leading yawners, consistent with yawning's link to testosterone.[25,26] Some investigators suggest that gaping yawns are threat gestures by alpha males, in which their formidable canine teeth are brandished before subordinates, whereas subordinates seldom yawn in the presence of dominant males.[25,27] This argument loses force because, at least in humans, true yawns are not under voluntary control and cannot be easily turned on or off to fit a circumstance. A behavioral psychologist would say that yawns make poor operants. However, the previously considered social inhibition of human yawning may be evidence that this primal social process has an involuntary basis, and individuals most resistant to its effects are alpha males and females. Being sexy beasts, we humans may be unique in that both sexes yawn equally often and are sexually receptive at all times.[28]

Yawning is triggered by androgens and oxytocin and is associated with other sex-related agents and acts.[25,26] In rats, most chemical agents that produce yawning and stretching also produce penile erection.[25,26] Although such antidepressant drugs as clomipramine (Anafranil) and fluoxetine (Prozac) typically depress sexual desire and performance, in some people they have the interesting side effect of producing yawns that trigger orgasms.[25] An opportunity exists for enterprising pharmacologists to move beyond building better erections to develop a drug that triggers orgasms in both sexes. Would such a biologically and psychologically potent agent be addictive? Having discovered a favorite new pastime, would enthusiasts never leave their homes, or eventually become jaded and indifferent? Although associated pleasure is undocumented, yawning and occasionally sexual arousal can be induced almost immediately in opiate addicts by the injection of naloxone or naltrexone, both potent opiate antagonists. The historical record yields more exotic but grim fare that links yawning and sexual arousal. Chroniclers of the late Middle Ages report yawning and ejaculation as symptoms of the terminal stages of rabies.[25]

Human yawners are not usually rewarded with orgasms, but yawning does feel good to most people, being rated 8.5 on a 10-point hedonic scale (1 = bad, 10 = good).[1] Given the similarities among sexual orgasm, yawning, and sneezing, including some resemblances among the characteristic facial expressions, it's perfectly reasonable to refer to the resolution of all three acts as "climax." Is the frustration of being unable to achieve sexual climax akin to the unsatisfying sensation of being stuck in mid-yawn or mid-sneeze? (The surprisingly sexy aspects of sneezing are covered in Chapter 7.)

The chronic urge but inability to yawn is quite disturbing to those who experience it, and several such people have contacted me about gaining relief. They probably go through life feeling "stuck in mid-yawn," as did my subjects in the clenched-teeth, nose-yawn, and open-eye yawn studies—a most unpleasant state. Would sexual climax offer them a measure of relief? I have also heard from people who seek relief from chronic yawning. An unusual case involved a child with state-specific yawning triggered exclusively by doing algebra problems. Cases of abnormal yawning are often confounded by multiple health problems and associated drugs. I may be the last resort for some people seeking knowledge about yawning. I can offer information but little else, collecting data that may be useful for those contacting me next, and possibly informing future research. Learning that you or your loved one is not the only person having an abnormal yawn condition is not much comfort, but it's something.

Spontaneous yawning and stretching have similar properties and may be performed together as parts of a global motor complex.[29] But they do not always co-occur—we usually yawn when we stretch, but we don't always stretch when we yawn, especially before bedtime. In other words, there are more stretchless yawns than yawnless stretches. (Voluntary stretches of the sort done when warming up for exercises seem not to be associated with concurrent yawns, although this proposition has not been empirically tested.) The yawn-stretch relationship confers a measure of yawn-mediated contagiousness to stretching—we often stretch while yawning contagiously. However, there is no evidence that yawnless stretches evoke contagious stretching and yawning in witnesses, unless they are cognitively associated with the hair-trigger stimulus of yawning. The yawn-stretch linkage starts early. J. I. P. de Vries and colleagues, then at the University of Groningen in the Netherlands, used ultrasound to trace yawning and stretching in the developing fetus.[23] They observed both yawning and synchronous yawning and stretching as early as the end of the first prenatal trimester (around eleven weeks post-conception).

The most extraordinary demonstration of the yawn-stretch linkage occurs in around 80 percent of people paralyzed on one side of their body (hemiplegia) because of a stroke. The prominent British neurologist Sir Francis Walshe noted in 1923 that when these hemiplegics yawn, they and their families are startled and mystified to observe that their otherwise paralyzed limb rises and flexes automatically in what neurologists term an "associated response."[30] These involuntary movements usually occur in association with yawning and less frequently with stretching, coughing, sneezing and laughing. In an earlier age, this response might have been evidence of demonic possession. Film enthusiasts may recall the disobedient arm of Dr. Strangelove. Yawning apparently activates undamaged, unconsciously controlled connections between the brain and the spinal cord motor system innervating the paralyzed limbs. It is not known whether the associated response is correlated with improvement, nor whether yawning is therapeutic for reinnervation or prevention of muscular atrophy. These individuals are alert and can probably catch yawns, but it is not reported if they do so.

Neurology patients with "locked-in" syndrome are alert but almost totally deprived of the ability to move voluntarily, whereas they can yawn spontaneously,[31] and probably contagiously. It will be informative to test the catchability of yawns by patients with reduced levels of consciousness. Contrary to popular opinion, individuals in a persistent vegetative state are not asleep and completely cut off from the world, and they usually have their eyes open. They seem unaware and unresponsive but can yawn spontaneously. Their susceptibility to observed yawns is unknown. Yawn contagion is a possible means of evaluating these uncommunicative individuals and measuring their residual brain function (sensory, motor, and social), level of consciousness, and, perhaps, prognosis for recovery.

The neural circuits for spontaneous yawning are located in the brain stem near respiratory and vasomotor centers, as suggested by the presence of yawning in anencephalic individuals born with only the medulla oblongata (brain stem).[32] Damage to these critical brain regions sufficient to alter yawning may be life-threatening. Contagious yawning relies on higher brain processes whose loss is less likely to be lethal.

The Folklore of Yawning

Having considered the motor act of yawning, let's test some of the folklore about when and why we yawn. Although often wrong, folklore poses interesting questions and is the repository of centuries of informal observations about human nature. A disadvantage of testing folklore is that when it's confirmed, you are accused of proving the obvious. Research has sometimes confirmed and extended common beliefs about yawning, but we have been rewarded with plenty of surprises. We can now contribute new, scientifically confirmed folklore.

We Yawn When Bored

Score a point for folklore. Bored people really do yawn a lot.[33] To induce ennui, study participants were asked to watch a television test pattern for thirty minutes, while those in the control condition got a thirty-minute dose of music videos (Fig. 1.4). However you feel about music videos, you will find them more interesting (less boring) than a test pattern of unchanging, colored, vertical bars. Participants yawned about 70 percent more during the test pattern than during the music videos. But yawning is not just for the bored. Anecdotal evidence reports yawning by Olympic athletes before their event, a famous violinist waiting to go onstage to perform a concerto, and dogs on the threshold of attack. The most dramatic such evidence was collected at my request by one of my students, an army special-forces soldier on academic posting. He observed the copious yawns of his fellow paratroopers before their first jump (Fig. 1.4). Bailing out of a perfectly good aircraft is not normal. These soldiers were certainly not bored.

The association between yawning and boredom is so widely known that faked, voluntary yawns have become a rude paralinguistic signal of boredom, inappropriate in polite society and in the courtroom of Will County, Illinois, circuit

figure 1.4 *When do we yawn? Some common beliefs have been supported by research. As conventional wisdom suggests, we yawn when we are bored, as when viewing a television test pattern. (A colored bar was used in the actual study.) But there is also anecdotal evidence of paratroopers yawning before making their first jump, musicians yawning while waiting to perform, and Olympians yawning before their event. (From Provine 2005)*

court judge Daniel Rozak.[34] The demonstrative yawn and stretch Clifton Williams made while his cousin Jason Mayfield was being sentenced for a felony drug charge earned him six months in jail for contempt of court, of which he will serve at least twenty-one days (*Chicago Tribune*, August 10, 2010). "I really can't believe I'm in jail," Williams said in a letter to his family. "I done set in this [expletive] a week so far for nothing." Judge Rozak obviously thinks otherwise.

We Yawn When Sleepy

As expected, participants recording their yawning and sleeping in a diary during a one-week period confirmed that we really do yawn most when sleepy,[29] especially during the hour after waking, and second most during the pre-bedtime hour (Fig. 1.5). A surprise came from accompanying data about stretching that participants also recorded in their diaries. After waking, subjects simultaneously yawned and stretched. But before bedtime, most participants only yawned. Look for this pattern in your own behavior. You can also observe this yawn-stretch linkage in your pet dog or cat as it arises from its slumber.

We Yawn Because of Too Much Carbon Dioxide or Too Little Oxygen

The legendary but unsupported factoid that we yawn when there is a high level of carbon dioxide or a shortage of oxygen in the blood or brain[35] is repeated so often that it has a life of its own, still being presented in the popular media and in medical school lectures. Yet the only test of this hypothesis, one that I conducted more than twenty years ago, soundly rejected them. Breathing levels of carbon dioxide a hundred or more times greater than the concentration in air (3 percent or 5 percent CO_2 versus the usual .03 percent CO_2) did not increase yawning, although participants did dramatically increase their breathing rate and tidal volume (deepness of breath), evidence that the gases had a major physiological effect. Furthermore, breathing 100 percent oxygen did not inhibit yawning. The effect of diminished oxygen was not tested because of danger to the participants.

Although both breathing and yawning involve respiratory acts and are produced by neurological motor programs,

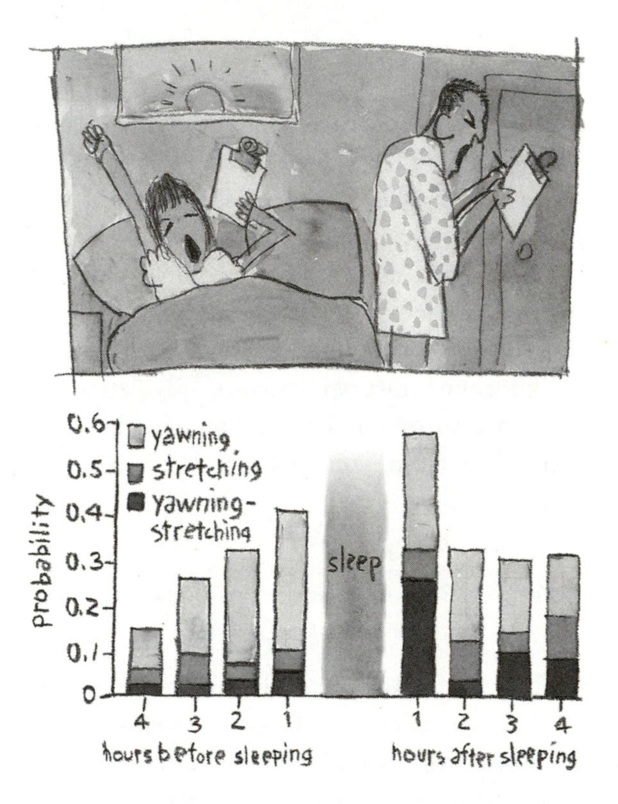

figure 1.5 *We yawn when we are sleepy, both before bedtime and after waking. Study participants recorded an increase in yawning as they approached bedtime; yawning was again common after waking. Yawning is often combined with stretching, especially after waking (upper), but not before bedtime. The graph (lower) shows the proportion of days in which yawns, stretches, and combined yawn-stretches were noted. (From Provine 2005)*

these programs are separate and can be modulated independently. Exercise, for example, that had participants huffing and puffing at high rates did not affect their rate of ongoing yawning. Test this proposition during your next jog by thinking about yawning.

The Enigma of Function

Does the flamboyant act of yawning, spontaneous or contagious, serve a function? Or is it much ado about nothing? Those expecting a single, definitive function may be frustrated by the answer to this question—it probably has many. Science reporters assigned a story about *the* function of yawning may be seeking an account that can never be told to their satisfaction. Yawning has many consequences and correlates that may serve as functions, including opening the Eustachian tube (clearing the ears), tearing, inflating the lungs, stretching, signaling drowsiness and boredom, marking the transition from sleep to wakefulness, and more.[36] These are facts. Picking one item from the list and proclaiming it *the* function neglects the diverse, system-wide impact of yawning. Further, yawning may be a relic of our developmental and evolutionary past. During embryonic development, for example, the gaping movement of yawning may sculpt the hinge of the jaw, or the ingress of amniotic fluid during a yawn may inflate the lungs and contribute to their development (Chapter 13). Yawning may have different functions for different species or developmental stages, each adapting the behavior to its needs. Proposed yawn functions range from the physiological (i.e., brain cooling, increased alertness) to the purely social (i.e., synchronization); samples of the literature on this are provided in the endnotes.[37] The options are reduced a bit by the testing and rejection of the previously considered hypothesis about carbon dioxide or oxygen.

I conclude with my own candidate for a yawn function that explains a wide range of behavioral and physiological data and accommodates a variety of other proposed functions. *Yawning is a response to and facilitator of change in behavioral or physiological state.* The change may be in sleepiness, arousal, aggression, brain temperature, or some yet unrecognized conditions—the times when we yawn most. Yawning and related stretching are massive neuromuscular and respiratory acts that stir up our physiology and promote these transitions. In humans, these changes extend to the group through contagion.

Consider the Bakairi people of central Brazil as observed by their first European visitor, nineteenth-century ethnologist Karl von den Steinen. Irenäus Eibl-Eibesfeldt quotes Steinen: "If they seemed to have had enough of all the talk, they began to yawn unabashedly and without placing their hands before their mouths. That the pleasant reflex was contagious could not be denied. One after the other got up and left until I remained with my dujour [diary]."[38]

These mindless behavioral echoes of yawning have unappreciated power. Contagious yawns propagating through a population drive a correlated ripple of physiology and emotion,[39] transforming group members into a collective super-organism. While the details of function are being settled, yawning provides fascinating phenomena, diagnostic tools, and insights into the neurological basis of social behavior. Yawning is evidence that beneath our veneer of culture, rationality, and language, ancient, unconscious processes influence our lives.

2

laughing

Consider the bizarre events of the 1962 outbreak of contagious laughter in Tanganyika (now Tanzania).[1–3] What began as an isolated fit of laughter in a group of twelve- to eighteen-year-old schoolgirls rapidly rose to epidemic proportions. Contagious laughter propagated from one individual to the next, eventually infecting adjacent communities. Like an influenza outbreak, the laughter epidemic was so severe that it required the closing of at least fourteen schools and afflicted about a thousand people. Fluctuating in intensity, it lasted for around two and a half years. A psychogenic, hysterical origin of the epidemic was established after excluding alternatives such as toxic reaction and encephalitis.

Laughter epidemics, big and small, are universal. Contagious laughter in some Pentecostal and related charismatic Christian churches is a kind of speaking in tongues (glossolalia), a sign that worshipers have been filled with the Holy Spirit ("Laughing for the Lord," *Time*, August 15, 1994).[4] Before looking askance at this practice, consider that it was present at the historic Cane Ridge (Kentucky) revival of 1801, and part of an exuberant religious tradition in which the Shakers actually shook and the Quakers quaked. Even John Wesley, founder of the Methodist Church, did some of his own quaking and shaking. Those experiencing the blessing of holy laughter spread it back to their home congregations, creating a national and international wave of contagious laughter. Contrast, now, the similarity between the propagation of such religious anointings and what was called the "laughing malady puzzle in Africa" (*New York Times*, August 8, 1963). They are strikingly similar, tap the same social trait, and are an extreme form of the commonplace, not pathology.

Laughter yoga, an innovation of Madan Kataria of Bombay, India, taps contagious laughter for his secular Laughing Clubs International.[5] The laugh clubbers gather in public places to engage in laughter exercises, seeking better fitness and a good time. Kataria's revelation was that only laughter is needed to stimulate laughter—no jokes are necessary. Meetings start with unison laughter exercises, moving on to more unusual variants. This self-described "laughing for no reason" produces real contagious laughter and is fun for the self-selected participants, but its claimed medicinal benefits remain conjecture.

The Tanganyikan and holy laughter epidemics, and laughter yoga, are dramatic examples of the infectious power of laughter, something that most of us may have experienced in more modest measure. Many readers will be familiar with the difficulty of extinguishing their own "laugh jags," fits of nearly uncontrollable laughter. We also share yuks with friends and join the communal chorus of audience laughter. Rather than dismissing contagious laughter as a behavioral curiosity, we should recognize it and other laugh-related phenomena as clues to broader and deeper issues. When we hear laughter, we become beasts of the herd, mindlessly laughing in turn, producing a behavioral chain reaction that sweeps through our group, creating a crescendo of jocularity or ridicule.

The use of laughter to evoke laughter is familiar to viewers of television situation comedy shows.[6] Laugh tracks (dubbed-in sounds of laughter) have accompanied many sitcoms since September 9, 1950. On that evening, *The Hank McCune Show*—a comedy about "a likeable blunderer, a devilish fellow who tries to cut corners only to find himself the sucker"—first used a laugh track to compensate for the absence of a live studio audience. Although the show was short-lived, the television industry discovered the power of canned laughter to evoke audience laughter.

The music recording industry earlier recognized the seductive power of laughter with the distribution of "The Okeh Laughing Record," which consisted of trumpet playing that was intermittently interrupted by highly infectious laughter.[7] Released shortly after World War I, it remains one of the most successful novelty records of all time. Acknowledging the commercial potential of this novelty market, jazz greats Louis Armstrong, Sidney Bechet, and Woody Herman, as well as virtuoso of funny music Spike Jones all attempted to cash in with laugh records of their own. Classicists may add that performers in the Athenian theater of Dionysus scooped everyone by more than two thousand years when they hired people to cheer or jeer to influence the audience and judges of their tragedy and comedy contests.

The innovation of laugh tracks in early television shows kindled the fears of some cold war era politicians that the pinko media was trying to surreptitiously control the masses. Psychology researchers jumped on the new phenomenon of "canned" laughter, confirming that laugh tracks do indeed increase audience laughter and the audience's rating of the humorousness of the comedy material, attributing the effect to sometimes baroque mechanisms (deindividuation, release restraint mediated by imitation, social facilitation, emergence of social norms, etc.). Decades later, we learned that the naked sound of laughter itself can evoke laughter—you don't need a joke.[8]

Prerecorded laughter produced by a "laugh box," a small battery-operated record player from a novelty store, was sufficient to trigger real laughter of my undergraduate students in a classroom setting.[8,9] On their first exposure to the laughter, nearly half of the students reported that they responded with laughter themselves. (More than 90 percent reported smiling on the first exposure.) However, the effectiveness of the stimulus declined with repetition. By the tenth exposure, about 75 percent of the students rated the laugh stimulus as "obnoxious," a reminder of the sometimes derisive nature of laughter, especially when repetitive and invariant.[10] With repeated exposure, I also grew to hate the sound of the canned laughter, wincing when curious students pushed the on button of one of the boxes in my office. Only disarmed boxes with batteries removed are now found on my desk. It is unpleasant to be the recipient of a scornful "ha."

Court fools, presidential aides, and corporate administrative assistants learn early in their careers that it is safer to laugh with the boss than at him or her. Plato and Aristotle correctly feared the power of laughter to undermine authority and lead to the overthrow of the state. Then as now, politicians' days are numbered when they become regular fare in comedy.

In our politically correct, feel-good, be-happy time, we are shielded from, and underestimate, the dark side of laughter that was better known to the ancients. If you think laughter is benign, be aware that laughter is present during the worst of atrocities, from murder, rape, and pillage in antiquity to the present. Laughter has been present at the entertainments of public executions and torture. On street corners around the world, laughing at the wrong person or at the wrong time can get you killed. The publication of cartoons of the Prophet Muhammad by a Danish newspaper triggered calls for the death of the cartoonists and a worldwide murderous rampage that left many dead and injured. Although radical Islam is most in the news, all monotheistic religions ruthlessly suppress humorous challenges to their spiritual franchise. The killers at Columbine High School in Littleton, Colorado, were laughing as they strolled through classrooms murdering their classmates ("Death Goes to School with Cold, Evil Laughter," *Denver Rocky Mountain News*, April 21, 1999). Laughter accompanies ethnic violence and insult, from Kosovo to Abu Ghraib prison in Iraq. Laughing *with* brings the pleasure of acceptance, in-group feeling, and bonding. But laughing *at* is jeering and ridicule, targeting outliers who look or act different, pounding down the nail that sticks up, shaping them up, or driving them away. Being laughed at can be a very serious, even dangerous business.

~

The ability of laughter alone to elicit audience laughter suggests that human beings have auditory "feature detectors," neural circuits that respond exclusively to this species-typical vocalization and trigger the neural circuits that generate the stereotyped action pattern of laughter[11] (Fig. 2.1). Have you ever been overcome by a comparable urge to echo the chant "hello-hello-hello"? The contagious laugh response is immediate and involuntary, involving the most direct communication possible between people—brain to brain. Contagious yawning (considered in the previous chapter) may involve a similar process in the visual domain. Detectors may also have evolved for the universal phonemic features of speech, but

figure 2.1 *You don't decide to laugh when you hear someone else laugh—it just happens. The ability of laughter alone to stimulate laughter in another individual suggests that human beings have an auditory feature detector, a neurological mechanism that responds specifically to this vocalization. In turn, this feature detector triggers other neural circuits that generate the stereotyped vocalization of laughter. The combined activity of the laugh detector and laugh generator is responsible for the unconsciously controlled act of contagious laughter. (From Provine 1996a)*

the variability and complexity of language and the absence of a contagious response to assay their activation make them difficult to discover and monitor.

If we have evolved a mechanism dedicated to the detection and replication of laughter, the acoustic structure of laughter is the key that unlocks the contagious response. We will now explore the act and sound of laughter.

Our lives are filled with laughter. From babyhood to old age, whether Greek or Sioux, we produce, hear, seek, and avoid this potent utterance in the universal human vocabulary.[12] Based on lifelong exposure, we should all be laughter experts, but we are not. Many of our assumptions about laughter are false, and these misconceptions color everything. One fallacious notion is that laughter is a mindful, voluntary act, and that laughter is a matter of speaking "ha-ha." Not so. Our inability to laugh on command invalidates traditional explanations about why we laugh and what it means, and makes it difficult to collect samples of laughter for study.[13]

Ask people to laugh and about half of them will claim that they can't laugh on command. The other half gamely honks out an obviously fake "ha-ha," proving the point. The degree of voluntary control can be evaluated systematically by using reaction times. It took over twice as long for participants in my study to laugh when told to do so (2.1 seconds) than to speak a similar "ha-ha" when asked to do so (0.9 seconds), but the true difference was probably even greater because many subjects produced what sounded like fake laughter (Behavioral Keyboard, in the appendix).

The longer reaction time for laughing (an involuntary act) than speaking "ha-ha" (a voluntary act) indicates that laughter is not spoken, and that different neurobehavioral mechanisms are involved.[14] Thus, none of us can accurately explain why we laugh. We are stubborn, arrogant beasts, unwilling to cede the illusion of self-control. Our explanations for laughing—to put someone at ease, to show friendship, to be playful, to ridicule, and so on—are simply post hoc rationalizations of the irrational that falsely presume the conscious control of involuntary acts. As we will see, there are better ways to explain laughter.

I captured a sample of laughter in the wild for acoustic analysis by venturing forth, microphone in hand, to meet idle students in the student union, visit staff in their offices, and intercept passersby, announcing, "I'm studying laughter. Will you laugh for me?"[15] This request was usually followed by bursts of genuine laughter, especially if I continued the banter and adopted a slightly goofy demeanor. With a sample of fifty-one cases of hard-won laughter, I retreated to a sound lab to analyze the data. Using a sound spectrograph, I was able to identify the acoustic signature of laughter.[16] I focused on classic ha-ha-type laughs, avoiding *laughspeak*, a kind of laugh/speech hybrid that is under more conscious control than ha-ha laughter and is often used by people to defuse a sensitive point.

Laughter has a stereotyped structure composed of short laugh syllables ("ha," "he," etc.) of about 1/15 second (75 milliseconds) duration that repeat at intervals of about 1/5 second (210 milliseconds), onset to onset (Fig. 2.2). Readers seeking more acoustic detail will appreciate that the sound spectrum of laugh syllables (notes) such as "ha" are composed of a stack of equally spaced frequencies that are multiples of a low, fundamental frequency, a sign of its strong harmonic structure. A syllable having a fundamental of 200 Hz, for example, would have harmonics of 400 Hz, 600 Hz, and so on. The syllables of women's laughter had an average fundamental frequency (502 Hz) almost twice as high as those of men (276 Hz), corresponding to women's higher-pitched voices. The short, periodic blasts of laugh syllables noted above often had vowel-like properties, but laughter sometimes can be noisy, lacking notable harmonic structure. The ha-has of laughter are not separated by silence. If you edit out "ha" and close the gaps, all that's left is a ghost of laughter, a breathy sigh.

The stereotypy of laughter is enforced by the mechanism of vocal production.[17] It's hard, in fact, to laugh in other than the usual way. Forget about acoustics for a moment and focus on vocal gymnastics. Consider three arbitrary laugh variants. It's easy to simulate "ha-ha-ha-ha," "he-he-he-he," and "ho-ho-ho-ho." Try it. Now try "ha-ha-he-he" and "he-he-ha-ha." Easy, right? Now try "ha-he-ha-he" and "he-ha-he-ha." That's tough, if you can do it at all. The point here is that constraints

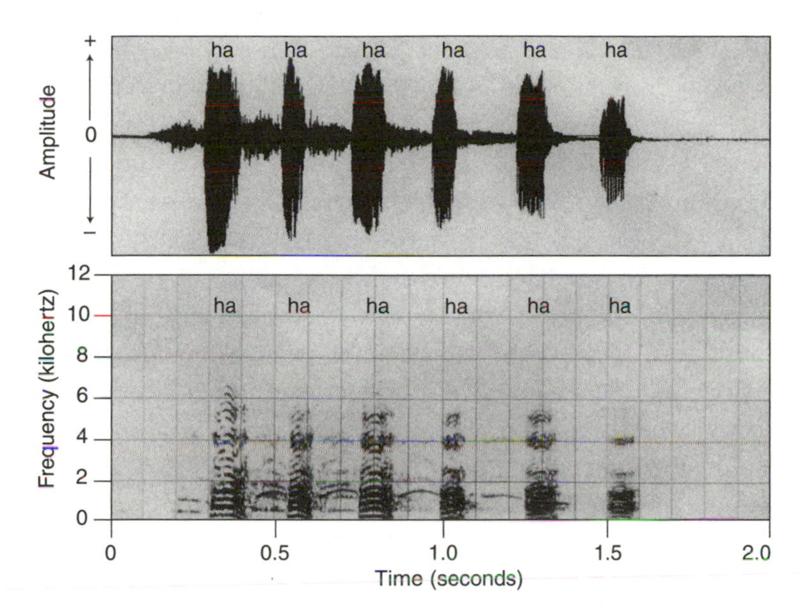

figure 2.2 *Characteristic features of laughter are evident in the regularity of the waveform* (upper) *and the frequency spectrum* (lower) *of a typical laugh (here consisting of six notes) of an adult male. The laugh notes ("ha") last for about 75 milliseconds (1/15 second) and repeat at intervals of about 210 milliseconds (about 1/5 second). In the frequency spectrum, each note features a stack of evenly spaced horizontal bands that are harmonics (multiples) of the note's fundamental frequency (the lowest band). Laughter is stereotyped but not fixed, varying around these central values. (From Provine 1996a)*

of your vocal apparatus force you to laugh in some patterns but not others.

Constraints also exist in the time structure of laughter. Try to laugh with very long intervals between laugh syllables, as in "ha——ha——ha" instead of "ha-ha." You can do it, but it doesn't feel or sound right. Now try to laugh at a faster than normal cadence, with a train of machine-gun-like, rapid-fire syllables: "ha-ha-ha-ha-ha-ha." Again, you may be able to do it, but it doesn't sound much like laughter, certainly not normal laughter. This time try to laugh with a series of super-short laugh syllables. Hard to do, and it sounds abnormal. Very long laugh syllables such as "haaaaa-haaaaa-haaaaa" are easier to perform but, again, don't sound normal. In other words, on the production side, the stereotypy of laughter is imposed by constraints of the vocal apparatus; on the perception side, it's constrained by the narrow range of vocalizations that sound like laughter. The universe of laugh-like sounds lies between these two anchorage points.

By focusing on the stereotypy of laughter, I'm not assuming the indefensible straw-man position of invariance.[18] I do suggest, however, that laughter has a strong central tendency, certainly when compared to speech. Indeed, if all laughter was different, we would not be able to identify the sound or evolve a detector for it. Individual and situational differences are certainly important, and there are probably identifiable acoustic nuances distinguishing laughter deemed "sly," "wicked," "ironic," "sardonic," or "gay," but those are problems for another day. It's best to start with core problems that are most easily solved.

~

Laughter is a universal human vocalization.[12] We all speak it. But as suggested by Aristotle more than two thousand years ago, are we unique in our ability to cackle, chortle, and guffaw? Are we *Homo ludens*, broadcasting ha-has into the void, listening for a response that may never come? Fortunately, this is a simpler problem than the search for alien intelligence and does not require the pricey hardware of NASA or Project SETI. You can consult a chimpanzee.

figure 2.3 *Play face of a young chimpanzee. The characteristic play face (mouth open, upper teeth covered, lower teeth exposed) accompanies the pant-like chimpanzee laughter. (From Provine 1996a, adapted from a photograph by Kim Bard)*

My revelations about laughter came while visiting some young chimpanzees (*Pan troglodytes*) at the Yerkes Primate Research Center in Atlanta. Kim Bard, then director of the nursery at Yerkes, and her assistant Kathy Gardner evoked laughter by tickling and cavorting with their young charges. I observed from outside a chain-link fence to prevent the young chimps from getting a death grip on the video gear.

When tickled and during rough-and-tumble, the chimps produced a "play face" (mouth open, upper teeth covered, lower teeth exposed)[19] and emitted the breathy pant-like sound that is characteristic of their laughter (Fig. 2.3). This breathy utterance is very unlike the sound of "chuckling" and "laughing," as Charles Darwin, Dian Fossey, Jane Goodall, and others have described ape laughter.[20] Context cues (tickle, rough-and-tumble, cheerful play face, playful demeanor) probably prompted these sophisticated observers to associate these panting ape vocalizations with the very different sound of human laughter.

To evaluate the sound of chimpanzee laughter for naive human observers, I played audiotapes of chimpanzee and human laughter to students in two of my college classes and asked students to identify on note cards what they were hearing.[21] Almost no one identified the chimp sound as laughter (2 of 119), whereas almost everyone recognized human laughter (117 of 119). In this auditory Rorschach test, the most common description of the chimp vocalization was "panting" (36 students)—most often believed to be that of a dog. Other descriptions included "asthma attack," "hyperventilation," "breathing problems," and "having sex." Some students (17) ascribed the chimp sound to nonbiological, mechanical acts, most commonly "sawing," but also "scraping," "erasing," "brushing," and "sanding." Breathy, panting, sometimes guttural chimpanzee laughter is obviously very different from its human equivalent. A person afflicted with chimpanzee-like laughter may have difficulty dating, getting tables at good restaurants, or running for high political office.

A visit to the sound lab revealed the acoustic differences between chimpanzee and human laughter (Fig. 2.4).[22] As noted previously, the vowel-like syllables of human laughter ("ha") have strong harmonic structure, last about 1/15 second, repeat every 1/5 second ("ha-ha"), and are performed by chopping a single expiration. Chimpanzee laughter, in contrast, is a noisy, panting vocalization that lacks obvious harmonic structure and is produced during *each* brief expiration and inspiration. This breathy, pant-pant laugh is the ancestral form. Human ha-ha laughter emerged sometime after our ancestral line diverged from that of chimpanzees about six million years ago.

The contrast between human and chimpanzee laughter brought a significant, unanticipated reward—the reason why we can speak and chimpanzees and other great apes cannot.[23] We laugh as we speak, by modulating an outward

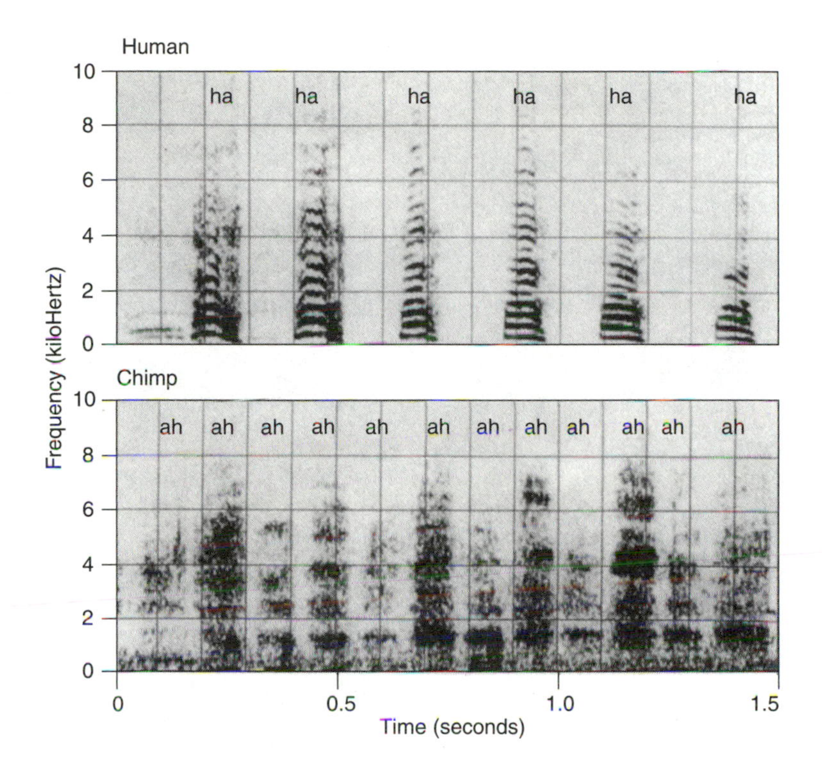

figure 2.4 *Frequency spectra of human* (upper) *and chimpanzee* (lower) *laughter are distinguished by the sharply defined onset and offset of the human laugh sounds. Noisy (breathy) chimpanzee laughter also lacks the clear harmonic structure of its human counterpart. (From Provine 1996a)*

breath (Fig. 2.5). If chimpanzees laugh as they speak, by producing one laugh sound per expiration and inspiration, we have identified an important and previously unrecognized constraint on the evolution of speech and language: breath control. The neck-up disciplines of linguistics and phonetics typically favor grammar and laryngeal mechanics and neglect such mundane neuromuscular matters, the nitty-gritty grunt work of sound production.

figure 2.5 *Human laughter and chimpanzee laughter differ in the coupling between laugh notes and respiration. The notes of human laughter, such as "ha," are produced by chopping a single expiration, a process similar to the production of speech. In contrast, chimpanzees produce only one laugh note, a breathy panting "ah," for every inspiration or expiration. The close coupling between breathing and vocalization in chimpanzees may partially explain the failed attempts to teach these animals to speak English. (From Provine 1996a)*

The vocalization of chimpanzees and other quadrupeds is captive to an inflexible neuromuscular system that synchronizes breathing and running (one breath per stride). Full lungs are necessary to brace the thorax for forelimb impacts during running. This is the reason that you hold your breath to lift a heavy weight. Without inflated lungs, the thorax is a floppy, air-filled bag. The evolution of bipedalism—walking and running upright on two legs—freed the thorax of its support function during locomotion and permitted flexibility in the coordination of breathing, running, and vocalizing.[24]

This is the basis of our *bipedal* ("walkie-talkie") *theory of speech evolution*. A bipedal human runner may employ a variety of strides per breath (4:1, 3:1, 5:2, 2:1, 3:2, or 1:1), with 2:1 being the most common.[25] With a vocal system in which utterances are no longer tightly linked to locomotion, the stage is set for the natural selection of speech and, incidentally, our species' characteristic "ha-ha" laugh. Lacking bipedality, the chimpanzee, especially the more vocally facile bonobo (pygmy chimpanzee), lingers on the brink of fluency, locked into a vocal system that permits only simple cries and calls.

If you threw a party for all laughing creatures, it would be a furry, warm-blooded, primate (maybe mammalian) affair, with lots of tickling, no jokes, and nontraditional hors d'oeuvres. The key is play; mammals do it, whereas reptiles and other creatures don't. Laughter is literally the ritualized sound of the labored breathing of physical play, the clearest case in the animal kingdom of how a vocalization evolved. *Laughter is primate onomatopoeia*. The origin of human laughter would have been discovered long ago if we pant-laughed like chimpanzees instead of being one step removed with our ha-has. Fortunately, this evolutionary saga has no missing link, and we can trace the evolutionary path.

The first step in human laughter evolution is the pant-pant of heavy breathing of physical play, followed by the ritualization of this sound in which pant-pant emerged as the vocal symbol for the act that produced it, followed by ha-ha, which is a human abstraction of the original pant-pant. The ritualized play vocalization is a signal that "I'm playing and not assaulting you." This signal is handy for us frisky, social primates. It announces the benign motive for our physical advances and prevents a defensive fist in the face or a knee in the groin. In human infants, laughter develops between three and four months of age and provides a means of communication with caregivers before the development of speech, signaling "Yes, continue," complementing the "Stop, you have gone too far" of fussing and crying.[26] Knowledge about laughter evolution eliminates a lot of hand waving and essay writing about fanciful origins and meaning of laughter and humor. Laughter derived from play; it's not a vocal contrivance to release tension, improve health, or acknowledge the wit of your dinner companion, at least not originally.

The roster of laughing creatures is growing, but certainty about what qualifies as laughter diminishes as we move from panting great apes to other mammals.

The most comprehensive work on the biobehavioral basis of play and associated vocalizations has been done by Jaak Panksepp and his colleagues.[27] If you have a rat that needs pleasuring, Panksepp is your man. His rats respond to light finger strokes of their ribs and belly (a tickle) with ultrasonic chirps (50 kHz), their play vocalizations. The chirps are also produced during rough-and-tumble with fellow rats. Although rats may lack a sense of humor, they certainly have a sense of fun. Contrast of the rat play chirps with primate laughter is limited by the lack of information about the sonic structure and means of production of the rat chirps. It will be interesting to learn if, like great apes, their play vocalization mimics the sound of labored breathing.

The evolutionary and comparative history of humor is a challenging topic because laughter, the gold standard of humor certification, is not present.[28] Without the response of laughter, we are left wondering about the intent of a beastly jokester. Inferences based on signing (American Sign Language) and other behavior suggest that ape humor, if it exists, resembles that of kindergarteners, based on intentional misnaming, misusing, or name-calling.

Roger Fouts, a prominent psychologist and primatologist, observed the famous signing chimpanzee Washoe using (or misusing) a toothbrush as a hairbrush. Moja, another of Fout's signing chimpanzees, called a purse a "shoe," put the purse on her foot and wore it as a shoe (misnaming and misusing). Francine "Penny" Patterson asked a signing gorilla named Koko to feed a baby doll with a bottle, but Koko held the bottle to the baby's eye instead of its mouth (misusing). In other instances, Koko treated rocks or other inedible substances as food (misusing), signed "flower" to a picture of a bird (misnaming), and signed "dirty toilet" for an offending caregiver. During his decades of experience, Roger Fouts never observed chimpanzee antics to be followed by laughter.

I forge recklessly into the paleohumorology fray, proposing my candidate for the most ancient joke—a feigned tickle. (Real tickling is disqualified because of its reflexive nature.) The "I'm going to get you" game of a threatened tickle is practiced by human beings worldwide and is the only joke that can be told equally well to a human baby or a chimpanzee. Both babies and chimps "get" this joke and laugh exuberantly. Until we know more, reports of animal humor, our own included, are highly speculative and may tell us more about the observer than the animal being observed

We move now from speculations about the animal psyche to the terra firma of human laughter. Laughter is a rich source of information about complex social relationships, if you know where to look. Learning to "read" laughter is particularly valuable because laughter is involuntary and hard to fake, providing an uncensored, honest account about what people really think about each other, and you.

~

Laughter is a decidedly social signal, not an egocentric expression of emotion. The social context of laughter was established by seventy-two student volunteers in my classes who recorded their own laughter, its time of occurrence, and social circumstance in small notebooks (laugh logs) during a one-week period.[29] Smiling and talking were also recorded, to provide contrasts with laughter and each other. The presence of media (television, radio, reading material, etc.) was noted because it serves as vicarious social stimulation. The sociality of laughter was striking. My logbook keepers laughed about *thirty times* more when they were around others than when they were alone—laughter almost disappeared among solitary subjects not exposed to media stimulation. Such a huge effect is gratifying in the social sciences, where effects are often tiny and variability high. Laughter provides many such big effects.

People are much more likely to smile or talk to themselves than they are to laugh when they are alone. Although we probably laugh or smile more when we are happy than sad, these acts are performed primarily in response to face-to-face encounters. You are least likely to laugh, smile, or talk immediately before bedtime and after waking, circumstances with reduced opportunities for social interaction. These data provide solid grounds for a behavioral prescription: if you want more laughter in your life, spend more time with other people. If no friends are physically present, you can dial them up on your phone. Even solitary television viewing may not be as socially impoverished as suggested by detractors and has something to offer the recluse: the people in the box. Better yet, if you are addicted to television, view it with friends.

Further clues about the social context of laughter came from the surreptitious observation of twelve hundred instances of conversational laughter by anonymous people in public places.[30] My colleagues and I noted the gender of the speaker and audience (listener), whether the speaker or the audience laughed, and what was said immediately before laughter occurred.

Contrary to expectation, most conversational laughter was *not* a response to jokes or humorous stories. Less than

figure 2.6 *Most human laughter takes place during ordinary conversations, rather than in response to structured attempts at humor, such as jokes or stories. The placement of laughter in conversation is striking and informative. A speaker typically laughs after a spoken phrase, such as "Where have you been? Ha-ha-ha," rather than in the midst of a phrase, such as "Where have . . . ha-ha-ha . . . you been?" The occurrence of laughter at the end of a phrase—the* punctuation effect*—suggests that a neurological process governs the placement of laughter in speech. The dominance of speech over laughter is suggested because laughter seldom interrupts the phrase structure of speech. (From Provine 1996a)*

20 percent of pre-laugh comments were remotely joke-like or humorous. Most laughter followed banal remarks such as "Look, it's Andre," "Are you sure?" and "It was nice meeting you too" (Fig. 2.6). Even our "greatest hits," the funniest of the twelve hundred pre-laugh comments were not necessarily howlers: "You don't have to drink, just buy us drinks," "She's got a sex disorder—she doesn't like sex," and "Do you date within your species?" Your life is filled with a laugh track to what must be the world's worst situation comedy. Mutual playfulness, in-group feeling, and positive emotional tone—not comedy—mark the social settings of most naturally occurring laughter. Laughter is more about relationships than humor.[31]

Another counterintuitive discovery was that the average speaker laughs about 46 percent more often than the audience. This contrasts with the scenario of stand-up comedy in which a nonlaughing speaker presents jokes to a laughing audience. Comedy performance proves an inadequate model for everyday conversational laughter. Analyses that focus only on audience behavior, a common approach, are obviously limited because they neglect the social nature of the laughing relationship.

The story became more provocative when we identified the gender of conversants in laughing relationships (Fig. 2.7). Gender determines the proportion of speaker and audience laughter. Whether they are speaker or audience (in mixed-sex groups), women laugh more often than men. In our sample of twelve hundred cases, female speakers laughed 127 percent more than their male audience. Neither males nor females laugh as much with female speakers as they do with male speakers, explaining the paucity of female comedians. On average, men are the best laugh getters. These differences are already present by the time when joking first appears, around six years of age. Based on this evidence, it is no surprise that your high school class clown was probably a male, a worldwide pattern.

Laughter is sexy. Women laughing at men are responding to more than their prowess in comedy. Women are attracted to men who make them laugh (i.e., "have a good sense of humor"), and men like women who laugh in their presence.[32] The next time you are at a party, use laughter as a guide to what people really feel about each other—and you. Laughter is a particularly informative measure of relationships because it is largely unplanned, uncensored, and hard to fake. Men and women mindlessly and predictably act out our species' biologic script. A man surrounded by attentive, laughing females is obviously doing something right, and he will comply

figure 2.7 *Speakers tend to laugh more often than their audiences. In a study of 1,200 episodes of naturally occurring laughter, a male speaker laughed somewhat more than his male audience* (upper left) *and a female speaker laughed somewhat more than her female audience* (upper right). *In contrast, a typical male speaker will laugh slightly less often than his female audience* (lower left). *The most striking differences between the sexes were found in episodes that involved female speakers and male audiences* (lower right), *when female speakers laughed more than twice as often as their male audience. (From Provine 1996a)*

by continuing to feed his admirers whatever triggers their laughter. Such good-humored fellows don't need a big supply of jokes. Their charisma carries the day. Laughter is not, however, a win-win signal for males and females; if it is used carelessly, you can laugh your way out of a relationship or a job.

The asymmetrical power of laughter and comedy for men and women is noted by comedian Susan Prekel, who bemoans that men in her audience will "find me repulsive, at least as a sexual being." In contrast, "male comics do very well with women."[33]

~

SM guy of the '90s, sensitive, humorous, communicative, 47 (looks 37), medium build, 5'5". Seeking a trim, healthy, pretty lady.

Cleveland Plain Dealer

Seabreeze dreams, desert life, bridging the gap. DWF, 54/5'4", seeking metaphysically aware man of keen mind, humor, inner resources.

San Diego Union-Tribune

Personal ads provide a direct approach to the value of laughter because people spell out their virtues and desires in black and white.[34] Laughter and humor are highly valued in the sexual marketplace. In 3,745 personal ads published by heterosexual males and females in eight U.S. national newspapers on Sunday, April 28, 1996, men offered "sense of humor" or its equivalent ("humorous"), and women requested it. Women, however, couldn't care less whether their ideal male partner laughs or not—they want a male who makes *them* laugh. Women sought laughter more than twice as often as they offered it. The behavioral economics of such bids and offers is consistent with the finding that men are attracted to women who laugh in their presence. Without such a balance between bids and offers, there would be no market for laughter and humor, and the currency of these behaviors would decline.

~

Amazingly, we somehow navigate in society, laughing at just the right times, while not consciously knowing what we are doing. Consider the placement of laughter in the speech stream.[35] Laughter does not occur randomly. In our sample of twelve hundred laughter episodes, the speaker and the audience seldom interrupted the phrase structure of speech with a ha-ha. (We did not study laughspeak.) Thus, a speaker may say "You are wearing that? . . . Ha-ha," but rarely "You are wearing . . . ha-ha . . . that?" The occurrence of laughter during pauses, at the end of phrases, and before and after statements and questions suggests that a lawful and probably neurologically based process governs the placement of laughter in speech. Speech is dominant over laughter because it has priority access to the single vocalization channel and laughter does not violate the integrity of phrase structure. The relationship between laughter and speech is akin to punctuation in written communication. I call it the *punctuation effect.*

The orderliness of the punctuation effect is striking because it's involuntary (we cannot laugh on command). If punctuation of speech by laughter seems unlikely, consider that breathing[36] and coughing also punctuate speech. Better yet, test the proposition of punctuation by examining the placement of laughter in conversation around you, focusing on the placement of ha-ha laughs. It's a good thing that time sharing by these airway maneuvers is neurologically orchestrated. How complicated would our lives be if we had to plan when to breathe, talk, and laugh? In other chapters, we learn that eating, drinking, crying, coughing, sneezing, belching, hiccupping, yawning, and vomiting also time-share in this busy facial orifice.

~

Our most recent research explored the sensory and social channels necessary for laughter among deaf signers and users of communication technology. The results confirm and extend the findings with face-to-face conversations.

In a research collaboration with Karen Emmorey at San Diego State University, we learned that the vocal laughter of deaf signers punctuates the stream of their signing (American Sign Language), is a response to mostly non-humorous material, and shows the same gender pattern as that of normally hearing vocal speakers, with deaf male signers being the best laugh getters.[37] Therefore, typical vocal laughter is maintained among individuals who can *see but not hear* their conversants.

Among 4,000 cell phone users, more laughter and smiles were produced by solitary phone users than by individuals sitting alone in public places, indicating that emotional expression is maintained with a medium that permits you to *hear but not see* your conversants.[38] Although this point is obvious to anyone who has used a telephone, in science it needed to be empirically confirmed.

The placement of emoticons (visual symbols of emotion) in text messages posted on Web user groups was used to study laughter and emotional expression in a symbolic visual medium, with such icons as "LOL" ("laughing out loud") and ☺ (the smiley face) being proxies for the real thing.[39] In a sample of one thousand emoticons found in Web messages, the most common were ☺ and "LOL." Consistent with our earlier findings with vocal conversations, emoticons usually followed banal comments (not jokes) and punctuated text. Emoticon use in text messages showed that typical patterns of symbolic laughter and emotional expression are maintained in a visual, linguistic medium in which you can *neither see nor hear* your conversant. We text like we speak and laugh.

~

I will close with a brief comment about what laughter *is not*—a medicinal procedure for body and soul.[40] The idea that laughter is good for us has become so pervasive that we neglect the fact that laughter, like speech and vocal crying (Chapter 4), is a vocalization that evolved to shape the behavior of other people. Laughter no more evolved to make us feel good or improve our health than walking evolved to promote cardiovascular fitness. Any health benefits of laughter, like those of walking or talking, are secondary, coincidental consequences of an act having another primary function. The popular media is full of frothy stories about "laughing your way to health" or "a laugh a day" produced by reporters who have been ordered by their editors to collect information for a story that their audience is presumed to want. Reporters comply. Praise of laughter, humor, and unbridled optimism is not balanced by stories of the costs of carelessness and a manic, Panglossian lifestyle, from reckless driving and drug abuse to bankruptcy.[41] Flapdoodle and snake oil are abundant, and honest accounting, positive or negative, is in short supply. Would we be having such conversations about the health benefits of reading, typing, or playing the kazoo?

Research on medicinal laughter is in a tentative, early phase and, like many other promising enterprises (e.g., genetic engineering, artificial intelligence, evolutionary medicine), will be punished for its early exuberance with a backlash of undue pessimism before a more accurate assessment is made of its true worth. My perspective about laughter/humor interventions is one of reserved optimism, and I suspect that some benefits presumed for laughter may be the product of the social setting of laughter (friends, family, lovers), not laughter itself. Some relief may be provided through placebo or distraction; call it entertainment and move on. Given the modest downside of laughter and humor, why not give it a try? It need not involve bedside assault by manic clowns or pom-pom-bearing cheerleaders for a positive lifestyle.

Whatever its cause and consequence, a life filled with laughter is clearly on the right track. The effects of laughter and humor will probably be varied, subtle, and complex, but we need not wait for physicians, biologists, or the FDA to give us permission to laugh.

Laughter feels good when we do it. Isn't that enough?

3

vocal crying

Think of life's most annoying sounds—those cringe-producing, fingernails-on-the-blackboard sounds that demand to be stopped, now! The crying of babies is near the top of the list of unpleasant sounds.[1] Whether it's that of your own child or of the ticking time bomb enthroned at the adjacent table at a restaurant, the cry grates on the senses, kicks in the doors of our auditory attention center, and demands action—to stop that damned sound. How unlike the laughter of the previous chapter, which makes you want to join in the merriment, unless the joke's on you.

Babies are mini-tyrants by necessity, weak, immobile beings who use their most potent weapon to motivate caregivers to provide services essential to life. Babies can't get things for themselves, cling to their mother's fur like other primates, or toddle along after their mother like ducklings. Crying is their "acoustical umbilical cord."[2] Given crying's critical nature, it's no accident that natural selection produced cries having maximum impact on caregivers. The sound of crying increases the breast temperature of lactating females[3] and triggers the milk letdown reflex.[4] Such coevolution of the sender's vocal process and the receiver's receptive process maximizes the chance that a cry is heard and the necessary services rendered.

Crying is innate, a part of our biological endowment that is needed immediately at birth; it is too important to relegate its acquisition to the slow and chancy process of learning. At birth and through infancy, a vigorous cry signals health,[5] and solicits a universal response from its maternal audience, who picks up the infant, puts it to the breast, and nurses it.[5,6] Because crying usually stops when an infant is picked up or held by a caregiver, crying is probably an attachment behavior.[6] However, infants do not always cease crying when picked up, as in inconsolable crying (i.e., colic).[7] Males seem an afterthought in this age-old, baby-mother ritual, but they too respond to a baby's cry, fumbling to find the off switch.

The traditional baby-mother relation may be glimpsed in the practice of hunter-gatherer societies (e.g., !Kung) in which the baby is usually within arm's reach, carried in arms or sling, on hip, chest or back, in contact with the mother or caregiver. Crying bouts in such nurturant cultures are shorter (but not less frequent) than expected by our standards, and the overall amount of crying and fussing may be reduced by up to 43 percent.[8] The lengthy crying bouts familiar in modern Western societies may be more an artifact of crib, playpen, and distance from an unresponsive caregiver than baby distress. The parental concern with "spoiling" noted by some pediatricians (e.g., Dr. Benjamin Spock)[9] seems misplaced and probably leads to more overall crying during infancy and beyond. Modern Western parents face the full force of their babies' frequent, unattenuated crying, a two-edged sword that can solicit caregiving or, in rare cases, abuse and death. The most common explanation by perpetrators in cases of shaken baby syndrome is to "stop the crying."[10]

Crying develops according to a universal schedule that transcends culture, parenting, and fleeting circumstance.[8] Present at birth, crying increases in frequency to a peak at around six weeks, then decreases until around four months, after which it remains at a steady level until the end of the first year.[11] Infants born about eight weeks prematurely show a similar, age-corrected peak in crying, indicating that this developmental pattern is due to maturation, not postnatal experience.[12] Early crying is diurnal, varying with time of day, with most crying occurring during the late afternoon and early evening hours,[13] just as most adult emotional tearing is concentrated in the evening hours.[14] Cultural and parenting differences may attenuate but not eliminate these basic underlying patterns.[8] Babies are highly predictable during the early months of life; they are awake, they sleep, and they cry.

During the first months after birth, crying is a noisy, untargeted demand for assistance, but it also can be triggered by hunger, pain, cold, or unknown causes. By the end of the first year, crying becomes more selective and nuanced, occurring when a caregiver is nearby, and reflecting social and environmental context, as is the case with speech, laughter, and other expressions of emotion.[15] This trend continues during the second year, when babies start to cry in response to strangers, separation from parents, and frustration (the last of which is a cause of temper tantrums).[16] Crying diminishes sharply during the second year, being displaced by speech as a mode of vocal communication. Why resort to a crude vocal instinct when you can specifically ask for what you want?

The most enigmatic cause of crying is colic, the term for intense bouts of inconsolable crying experienced by up to 40 percent of babies, and the most common clinical complaint by mothers to pediatricians.[8] Colic is typically defined in terms of Wessel's "rule of threes," in which an infant cries for more than three hours per day, for more than three days per week, for more than three weeks.[17] Despite their apparent anguish, most colicky babies do not exhibit other signs of pathology, including the common folk diagnosis of "gas pains." Colic may be only a more intense form of normal crying, perhaps being an expression of vigor, not pathology.[5]

All for One, and One for All: Contagious Crying

If one cry is not stressful enough to caregivers, imagine a baleful chorus of bawling babies. Even newborns are programmed to cry or produce a stress reaction (vocal, facial, or physiological) when exposed to a crying infant.[18–22] Forty-three years passed between the seminal report of contagious infant crying by Charlotte Buhler and Hildegard Hetzer in 1928 and the next systematic investigation, that of Marvin Simner in 1971, but interest in the topic is increasing due to the presumed relevance of contagion to issues from empathy to mirror neurons.

Contagious crying and associated stress reactions are probably innate because they are detected in the early days of postnatal life.[19,21,22] Elements of the response extend through the first year[20] and into later childhood.[23] Newborns can discriminate between recordings of their own cries and those of other infants, showing more distress when hearing cries not their own[21,22] or of a different species (chimpanzee).[21]

The literature about contagious crying is a specialty unto itself, almost devoid of references to other, obviously contagious acts such as yawning and laughing. The neglect is reciprocated in the literatures of yawning and laughing, forcing scholars to build their own, tentative bridges between contagious behaviors and leading them to wonder if the exercise makes sense or is worth the effort.

Having reviewed the general significance of contagious behavior in previous chapters about yawning (Chapter 1) and laughing (Chapter 2), I'll cut the preliminaries and immediately point out a unique feature of contagious crying—its precociousness. Spontaneous crying is present at birth, with contagious crying present at birth or shortly after. Yawning, while present at birth, does not become contagious until several years later. Laughter, in contrast, does not develop until three to four months after birth, and the age of onset of its contagiousness is unknown. These apparently arcane developmental milestones become salient if contagious crying is presumed to be a precursor of emotional sharing and empathy, as it is by some developmentalists.[20,24] Important questions remain unanswered.

Why is contagious crying any more relevant to the development of empathy than contagious yawning, which does not mature until several years later, or laughter? How are various contagious acts related? Is there a difference in the contagiousness of yawns and laughs if produced by self or other, as with cries? Does each contagious act bring its own modest measure of empathy, pooling resources until critical mass is reached and true empathy somehow bursts forth in full flower? These issues are necessarily speculative, but they are central. Contagious acts may be primal, modular, lower-level neurobehavioral mechanisms for social and physiological coupling, each with its own evolutionary and developmental history and purpose. They may not be precursors of anything. Empathy and other forms of emotional sharing involve more global, complex processes operating at a higher level of brain function and conscious awareness, and may not be built on a foundation of contagious crying, laughing, and yawning.

~

To provide a developmental endpoint for crying, we jump now from babies to adults, a transition that reflects the relative amount of scientific literature. A lot more is known about crying during the first year, and to a lesser degree during adulthood, than during the ages in between. Big developmental changes are involved, including the shift from noisy, vocal crying to the silent, visual cue of tearing, a transformation beginning during the two- to four-month period when vocal crying starts to decline and emotional tears appear (Chapter 4). As an adult, you are mobile and self-sufficient, can use speech to request aid, and need not broadcast a vocal alert to solicit help.

Reflection on your own behavior provides a reasonable account of the most obvious developmental trends. As an adult, you cry much less than when young, and your crying is more often subdued, teary weeping than the demonstrative, vocal sobbing of childhood. You may also recall that the trauma that causes your crying is now more often emotional than physical. However, whether intentional or not, as adult or child, you cry to solicit assistance, whether physical aid or emotional solace. Paradoxically, your adult cry for help is more private than the noisy, promiscuous pronouncement of childhood, often occurring at home, where it finds a select audience. The developmental shift from vocal crying to visual tearing favors the face-to-face encounters of an intimate setting (Chapter 4). The maturation of inhibitory control gives adults the ability to select where and when crying occurs, or to inhibit it altogether, options less available to children.

Crying, as considered so far, explores a path well worn by those struggling to decode a primitive vocal clue about the needs, health, and psyche of fellow *Homo sapiens*. We now move down a less traditional path, turning over an occasional stone to see what lies beneath.

Crying versus Laughing

Crying and laughing, like joy and despair, and the ancient Greek masks of tragedy and comedy, are not chance pairings, but reflect the duality of these complementary behaviors and their neurological mechanisms. Learning about one informs us about the other.

Crying and laughing are both universal (species-typical) human vocalizations that develop without benefit of learning, as indicated by their presence in blind and deaf children, who can neither see nor hear the acts that could be modeled.[25] Crying wins the developmental sweepstakes, being present at birth, whereas laughing lags behind, emerging three or four months later.[26] In fact, crying develops even earlier than the full-term birth date, arising in premature infants as young as twenty-four weeks of gestational age.[27] Crying must develop early, as the life-and-death demand for essential caregiving, unlike the less critical social link of laughing. The earlier development of crying relative to laughing is a measure of its adaptive significance and phylogenetic antiquity—the most ancient behaviors tend to develop first. In this context, it's significant that the recently evolved emotional tearing component of crying emerges months after the more ancient vocal cry.

Specialists may argue whether there is a typical cry or laugh, but enough is known about these vocalizations to provide vivid contrasts. A cry is a *sustained*, voiced utterance, usually of around one second or more (reports vary), the duration of an outward breath. Think of a baby's "waaa." (Phoneticians may offer more precise transcriptions, but "waaa" is in common usage and serves our purpose.) Cries repeat at intervals of about one second, roughly the duration of one respiratory cycle. Again, reported values vary, depending on criteria and developmental age. A laugh, in contrast, is a *chopped* (not sustained), usually voiced exhalation, as in "ha-ha-ha," in which each syllable ("ha") lasts about 1/15 second and repeats every 1/5 second.[28,29]

Everything about laughing is much shorter in duration (faster) than crying. The staccato sounds of laughter are about fifteen times shorter than cries and repeat at a rate that is about five times as fast. (These multiples are necessarily rough approximations, but the essential issue is the huge size and direction of differences.) Laughs and cries, like speech, are both produced during an outward breath, whether the parsed, rapid-fire blasts of "ha-ha" or the sustained wail of crying. The modulated exhalation essential for laughing requires a level of vocal control absent at birth and similar to that necessary for speech.[29,30] At an undetermined time during development, a sobbing form of crying ("boo-hoo-hoo," etc.) emerges, which has a faster, cyclic structure more like that of laughter (and speech).

Our vocal apparatus enforces these cadences, especially for laughter. As noted in the previous chapter, it's difficult to laugh using unusually long or short syllables, or at longer or shorter intersyllabic intervals, and the result usually sounds very un-laugh-like.[29] It's much easier to vary the duration and repetition of the graded, vowel-like cry than the explosive laugh. Try it. Given these constraints, it's understandable that there is much more variability in the duration and cyclicity of crying than laughing.

Crying and laughing both show strong *perseveration*, the tendency to maintain a behavior once it has started. These acts don't have an on-off switch, a trait responsible for some quirks of human behavior. Whether baby or adult, it's easier to prevent a bout of crying than to stop it once under way. Crying causes more crying. Likewise, laughter causes more laughter, a reason why headliners at comedy clubs want other performers to warm up the audience, and why you may be immobilized by a laughing fit that can't be quelled by heroic attempts at self-control.[29] In fact, voluntary control has little to do with starting or stopping most crying or laughing.

Using reaction time, our research team determined that both crying and laughing are under weak voluntary control, corresponding to keys toward the left, hard-to-play side of the Behavioral Keyboard (in the appendix). When subjects were asked to cry, reaction time was the longest (9.8 seconds) of any behavior that we measured. Only 3 of 103 subjects (3 percent) even attempted to cry during the 10-second time limit. Laughter had a shorter reaction time, 2.1 seconds, and 93 of 103 participants (93 percent) responded, though many of these utterances may not have been true laughs. Although the latency for laughing is much shorter than that for crying, it's over twice as long as the 0.9 second required to utter "ha-ha," a simulated laugh. (A simulated crying sound was not measured.) Clearly, crying and laughing are under weak voluntary control and are not produced with the facility of speech. This inaccessibility is the motive for the "method" technique, which teaches actors to take an indirect path to creating emotions by imagining the situation that produces them.

Our brains conspire to multiply crying and laughing by perseveration at the level of the individual and contagion at the level of the group. Readers are familiar with the *contagion* of laughing[29,31] and yawning[32,33]—the tendency to involuntarily replicate these acts when they are observed, as detailed in the previous chapters about laughing and yawning. Contagious crying, discussed above, is less well known but is responsible for the phenomenon of cries sweeping through newborn nurseries. There are no "crying tracks" equivalent to the laugh tracks of television situation comedies. The sobs and tears of the actors may be sufficient to communicate misery and grief; no technological Greek chorus of wailing is needed.

In complementary ways, both crying and laughing, contagious and not, are social vocalizations. Crying is the solicitation of caregiving and succor, with its predominant stimulus shifting gradually from the physical injury of childhood to the emotional trauma of adulthood. Although its cause may be social, adults engage in solitary crying. Laughter, in contrast, is about social play and bonding. With increasing age, the setting of laughter shifts from the physicality of tickling and rough-and-tumble to the cognitive and linguistic arena of conversation, but its stimulus is always another person, whether physically present, in your mind, or a virtual representation via radio, television, telephone, or text message (paper or electronic).

Crying is most frequent during childhood, especially during infancy, in contrast to laughter, which is maintained at a relatively high level throughout life. (Children probably laugh more than adults, a result of their greater amount of play, but I have not found the source of the often quoted relative frequencies of child and adult laughter.) You probably laugh on a daily basis, whereas you may have to search your memory for your last crying episode. The social rewards for laughter are certainly greater. Whether in professional life or on the social circuit, people are attracted to those of good humor who spread cheer and a sense of play. Weepers are socially and psychologically more expensive than laughers, creating unease by covertly requesting services that others, however compassionate, may not want to provide. Chronic weepers really are needy, high-maintenance individuals. Crying is socially costly; laughter is cheap.

Our contrast of crying with laughing concludes with psychopathology and neuropathology, the study of things that go wrong with the brain and behavior. This is a fitting finale because inappropriate affect—crying and laughing in the wrong setting, or in an atypical manner—is one of the most common symptoms of pathology, and a cause of psychological trauma and social estrangement for patient, family, and friends.

As Oliver Sacks taught us in his neurological case studies, the most important lesson about the abnormal may be what it teaches us about the normal, about being human. We will explore this theme with a thought experiment.

Could an extraterrestrial pass a test for humanhood? After a cursory appearance and sniff test, inquisitors might examine crying and laughing, acts that are irrational, involuntary, and socially complex in ways not easily understood or learned by Mr. Spock of *Star Trek*, a replicant in *Blade Runner*, or the computer Hal in *2001: A Space Odyssey*. Laughing and crying are harder to fake than etiquette, table manners, and golf, screening tests for social class.

Even humans sometimes fail the laugh/cry test, earning unwelcome attention, social estrangement, and hostility.[29,34] Uncontrolled emotional outbursts in some victims of forebrain damage or neurological disease are an unrecognized barrier when those individuals wish to return to the workplace. Inappropriate laughing has particularly severe consequences, being interpreted as ridicule, disrespect, or craziness. Whether the product of a normal or damaged brain, it can get you fired, divorced, institutionalized, or even killed. At best, abnormal laughter seems creepy, weird, or diabolical. The social costs of inappropriate crying are lower: errant sobs and tears may inhibit aggression and solicit caregiving and sympathy, but they can also diminish social rank and perceived competence. No one is threatened by weeping. It's just pathetic.

Neurological disorders that involve vocal crying and laughter are numerous and will grow as clinicians become more attentive of symptoms and significance, moving beyond such global categories as "abnormal" and "inappropriate affect" to specify exactly what is abnormal or inappropriate. A sample of pathological crying and laughing[34] of varying rarity includes epileptic seizures that produce crying (dacrystic seizures) or laughter (gelastic seizures) but not both; Rett disorder, with excessive laughing, crying, and irritability, affecting only females; and Wilson disease (hepatolenticular degeneration), a genetically based degenerative disorder of copper metabolism that produces both involuntary crying and laughing. By far the most common cause is Alzheimer disease, a degenerative brain disorder, with about 39 percent of affected individuals showing pathological affect, 25 percent showing crying episodes, and 14 percent showing laughing or mixed laughing and crying episodes.

The emotional lability (instability) present in some people with multiple sclerosis (MS) and amyotrophic lateral sclerosis (ALS, Lou Gehrig's disease) is manifest in disproportionate, uncontrollable and embarrassing emotional outbursts—inconsolable crying in response to a sad news story, or joyous laughter in response to meeting an old friend. The experience is like being a competent driver of a runaway car that has no brakes and perversely overresponds to every tap of the accelerator or turn of the wheel, swerving from one emotional extreme to the other. Such expressive swings are a bit like our own, but much stronger and longer in duration, an example of the pathological as extreme variant of the norm.

Consider excerpts from Frances McGill's remarkable account of her life with ALS.[35] Her keen intellect is spared as she progressively becomes unable to move and control vital body functions.

> *Episodes of laughter and tears may have slight connection with my actual frame of mind. . . . I begin to smile, but the smile becomes an exaggerated grin. . . . If I yield to the impulse, it becomes the onset to equally uncontrollable giggles, which in turn so embarrass me that I become angry, humiliated and subject to uncontrollable tears—it is a vicious, see-sawing circle.*
>
> *That which in another day and age, might have been a slight misting of the eyes or lump in the throat, is now . . . the screaming monster of a "bawl." . . . They destroy me—they weaken and crumble me . . . those deep debilitating agonizing episodes. They are not gentle rain. They are more hurricanes.*

Her story dramatically illustrates the necessity of distinguishing between the externally observable behavior of laughing and crying and subjective mood states—her hyperactive emotional displays were not associated with a mood disorder, like those found in depression or mania. However, more suggestible ALS patients may be convinced by their excessive crying and family reaction that they really are depressed or losing their sanity, a self-fulfilling prophecy leading to inappropriate medication and making life even more difficult to bear. Physicians may neglect inappropriate emotionality because patients are unlikely to die from laughing and crying (although death by laughter has happened on rare occasions), and patients may underreport these symptoms because of social stigma or presumed irrelevance. Facing an apparently untreatable disease, physicians often forget that drug treatment to eliminate pathological emoting may be only palliative but can greatly improve patient self-esteem and social relationships. And a patient's emotional outbursts do not happen in a vacuum. Treating the patient also benefits family, friends, and caregivers.

The stimulus for laughing and crying, whether normal or pathological, is usually another person, who is either physically present or "in your head" via thoughts or imagery. Emotionally labile people may use this knowledge to devise strategies to dampen a pending outburst, such as social disengagement (for example, looking away from a person and staring at the floor) or cognitive disengagement (such as concentrating on a chair to block emotional thoughts).

At a more philosophical level, happiness and sadness, like life's other ups and downs, are relative judgments and subject to swings large and small. In the economics of the human psyche, the price of an emotional high, whether the product of life experience, drugs, or pathology, is the low that follows, and vice versa. This is a manifestation of an eternal truth: "There ain't no such thing as a free lunch." Yogis recognize this inescapable truth, offering a path for inner calm and the good life that may not be for every taste: sail as a ship on serene waters, avoiding the peaks and troughs of rough emotional seas. I offer my own, less ascetic path: "Moderation in all things, including moderation."

4

emotional tearing

What's so remarkable about tears? Tears bathe, lubricate, and heal the eyes,[1–4] but is that cause for excitement? For the physiologically inclined, it's notable that tears contain lysozyme, the body's own antibiotic—who wants mossy eyeballs teaming with microbes? Tears' contribution becomes obvious when they cease, producing the discomfort and pathology of dry eye. Basal tears are continuously secreted to lubricate and wet the eye, improve optical performance by smoothing the otherwise rough corneal surface, and are a reflexive response to irritation (e.g., abrasion, onion). Tears form a multilayered film on the eye, an inner layer with lubricating mucin, a watery middle layer, and an oily outer layer that reduces evaporation and drying. Important points all, but more revelation than revolution.

The revolution is *emotional tearing*, the uniquely human secretion of tears from the lachrimal glands as a visual signal of emotion.[5,6] Emotional tearing is a transformational event in our evolution as a social species. The production of emotional tears is so central to our identity that it has become a criterion for membership in the elite club of sentient beings. The high stakes for membership are recognized by animal rights activists and reflected in such books as *When Elephants Weep: The Emotional Lives of Animals*.[7] However, the dispassionate evaluation of evidence indicates that neither elephants nor chimpanzees, our primate cousins, shed an emotional tear. The exclusivity of humankind's crown jewels—language, laughter, and tool use—has been challenged, but emotional tearing still stands as a uniquely human trait. Even human newborns don't gain membership in the emotional tear club until several weeks or months after birth.[8]

Emotional tears are a universally understood and uniquely human signal of sadness and other emotional states and acts, including vocal crying, grief, despair, pain, happiness, anger, and empathy, as well as yawning, laughing, and sneezing. The association of tears with sadness is so strong as to be a cliché, the bedrock of folk wisdom about our species. But is it true? Do tears really signal sadness?

Remarkably, a search of the scientific literature about crying and tearing revealed no test of this tacit assumption. Published reports on tearing are full of interesting facts and findings about physiology, gender, personality, social context, culture, psychopathology, and health.[3,6] There is much opinion but little empirical support for the popular notion that emotional tears elevate mood.[9] I encountered the provocative proposal that emotional tears selectively secrete hormones and other substances that may provide cathartic relief and shed toxins. But my enthusiasm was diminished by the realization of the minute volume of tearing per crying bout (about 1 ml).[1] If tears reduce stress, drooling and urination may be cathartic Niagaras. There are a lot of data, but not what I was looking for. The large and energetic cadre of emotion researchers who study everything from face perception to psychophysiology somehow overlooked the problem of tears as visual signal.

My small team of mostly undergraduate psychology research students and I embarked on the examination of tears as a visual signal. Our sole physicist bailed, unpersuaded by my explanation that this was applied physics. The novelty of our problem brought the excitement of the chase but also the opportunity for major error. We started at what seemed to be the beginning.

The role of tears as a signal of sadness was examined by contrasting the perceived sadness of facial images with tears with copies of those images that had had the tears digitally removed with Adobe Photoshop.[5] Two hundred images of faces—fifty with tears, fifty with tears removed, and one hundred tear-free control images—were presented sequentially as a slideshow, in counterbalanced order, on a computer monitor. Our eighty experimental participants had up to five seconds to rate each image on a 7-point sadness scale—from 1 (not sad at all) to 7 (extremely sad)—before the next image appeared.

As expected, facial images with tears were rated as sadder in appearance than the same images with tears removed. (You can approximate the effect of tear removal by using your finger to block out tears in a photograph.) We term the ability of tears to make faces look sadder, the *tear effect*. Big deal! We confirmed the obvious: that *tears contribute to perceived sadness*. If this test of the expected hardly seems worth the effort, recall that our previously considered studies of yawning and laughing found numerous errors in folk wisdom and produced serendipitous findings. Our good fortune continued.

A startling but incidental discovery was that *tears resolve ambiguity of facial expression*.[5] The removal of tears often produced faces of uncertain emotional valence—perhaps awe, concern, contemplation, fright, or puzzlement, not simply less sadness. In other words, without tears, faces may not appear very sad, especially if they fall in the middle range of the emotional spectrum. This is a more significant finding than it seems on first hearing. Emotional tearing taps an already present secretory process to add nuance and range to the limited neuromuscular instrument of facial expression and contributes to the emergence of *Homo sapiens* as a social species. Tears are not merely a benign correlate of sadness.

In the spirit of artist Roy Lichtenstein, who produced many cartoons of crying women, I offer a cartoon demonstration of the tear effect developed with colleague and graphic designer Erin Ouslander (Fig. 4.1). (The study used halftone color photos of actual people with and without tears, not cartoons.) The ability of simple cartoons to portray the tear effect is a measure of its robustness. Note how the tearful face in the upper panel looks sadder than the middle image, which is identical except that the tears were removed. If you are skeptical about our graphic manipulation, confirm it by using your finger to block out the tears in the upper image.

Erin and I succumbed to the temptations of Photoshop and generated the fanciful, lower panel, answering a question that was not considered in our study and somehow escaped the attention of the world's tear researchers: do tears flowing upward have an emotional impact? We illustrate that the tear effect works only when the tears are flowing downward, not upward. The practical implications of answering this question that no one had asked: for maximum emotional effect, don't cry while standing on your head. The upward tears do not even look tear-like. A NASA grant will be required to explore the appearance and signaling capacity of tears in zero gravity. Unable to drain normally, tears would probably pool in the eye, held by surface tension, clouding vision unless blotted.

figure 4.1 *When tears are removed from a crying face, the resulting tearless face appears not only less sad but emotionally ambivalent, perhaps expressing awe, puzzlement, or concern* (center). *Tears add emotional valence to the limited neuromuscular instrument of the face; this is the* tear effect. *The mere presence of tears is an ineffective visual cue; if tears run upward instead of downward from the eye* (lower), *they lose their emotional impact. (Designed by Erin Ouslander)*

More work must be done to understand the full contribution of tears to the perception of sadness and other emotional states, including their blends. Do tears, for example, make a person appear more needy, helpless, frustrated, or powerless, as well as sadder? Do tears amplify a perceived emotional expression, add a unique message, or contribute a subtle nuance interpreted as sincerity or wistfulness? Do tears express a blend of emotions, such as anger and powerlessness? Does a happy face with tears appear more or less joyous, or something in between, perhaps described as "bittersweet"? Is a tear a secretory exclamation point that adds power to any expression? Are tears more prominent or emotionally potent on dark skin than on light skin? Does the race, sex, or age of a tearful face influence ratings of sadness or other emotions? (The current study detected no difference between sadness ratings by adult men and women, or by adult raters of different ages.) Are there interactions between the race, sex, and age of sender and recipient of tearful signals, such that, for example, children seem sadder than adults, women seem sadder than men, or same-race individuals seem sadder than individuals of different race?

My conclusion about tears as emotional stimuli received unanticipated confirmation from a student attendee at a scientific meeting where I first presented this study.[5] She told me that a medical condition in previous years deprived her of both the ability to tear and to clearly express emotion. Absent teary eyes and cheeks, her friends and family could not comprehend her occasional bouts of sadness, and she was forced to explain her distress, in a shaky voice, at the times when it was most difficult. Her available cues of facial expression were often inadequate. She missed the automatic, potent, and informative signal of tears present before her illness. Her story was persuasive and prompted plans for research about the effect of dry eye syndrome on emotional communication.

The inability to produce emotional tears is also a concern of some depressed patients being treated with serotonin reuptake inhibitors (SRIs), which rapidly improve the symptoms of excessive or inappropriate crying.[10] One patient who was an emotional person was distressed that she could no longer weep in church or while watching movies. Another complained that she was unable to weep during movies or during arguments with her husband; her only other change in emotional responsivity was the inability to achieve orgasm. A third patient complained of an inability to obtain emotional release. Being unable to grieve the death of her pet dog, she stopped taking medication and one day later was able to cry when thinking about her dead pet. She later resumed taking the drug and her tearing again disappeared.

Individuals who never developed tearing or have lost tearing because of drug treatment or disease provide a unique "experiment of nature," the opportunity to use a naturally occurring condition to test a basic issue about the role of tears in emotional signaling. People with dry eye, especially if highly emotional, may share the experiences of the student at the scientific meeting and the psychiatric patients. This problem remains open because it lies at the intersection of disciplines. Ophthalmologists probably focus on the immediate problems of tear production and replacement, and patients probably don't complain about what they presume to be a psychological quibble of little medical consequence.[2] Psychiatrists are more concerned with stopping pathological crying than with understanding the consequence of doing so. It's hardly surprising that emotion researchers who neglected the effects of tearing in normal subjects would attend to their cessation in drug treatment of dry eye.

Now that you are thinking about a life without tears, consider that you were born that way. Unlike vocal crying, which is present at birth, emotional tearing does not appear until several weeks to months later. This is a big deal. Although this developmental lag has been known for many years (e.g., by Darwin in 1872), its consequences are not appreciated by the legions of researchers who study babies. The emergence of tearing opens a new channel of visually mediated emotional communication between infant and caregiver. Tearing nicely complements vocal crying as a signal, and appears when vocal crying is declining. Tearing, as a visual signal, is silent and requires facial illumination of the sender and line-of-sight contact by an observer. Vocal crying, in contrast, works in darkness or around obstructions and reveals the location of the sender. Taken together, visual tearing and vocal crying provide a versatile, multimodal link with caregivers in childhood and companions in adulthood. Emotional tears may also have a previously unrecognized chemosensory dimension.

Recent research from the laboratory of Noam Sobel at the Weizmann Institute in Israel suggests that emotional tears are a chemical as well as visual signal and can work even in darkness.[11] In several studies, Sobel and colleagues found that men who sniffed drops of women's emotional tears (triggered by watching sad movies) in a pad beneath their noses were less sexually aroused than when they sniffed a saline control solution that had been trickled down women's cheeks. Evidence of depressed male arousal was found in skin responses, brain imaging, lowered testosterone levels, and lower self-reported sexual interest. In one study, for example, tear-sniffing men rated women in photographs as less sexually attractive than when sniffing saline. Only the effect of women's tears on men was examined in this preliminary study; the effect of women's tears on other women and of men's tears on women and other men was not examined.

The behavioral significance of these findings is necessarily speculative. Are women's emotional tears a primal, nonverbal means of saying no? Are tears a signal of lowered fertility, perhaps associated with menstruation, when women may be more tearful? Tears may also have different effects in different contexts. The tears of sadness may, for example, have different effects than those of happiness, coughing, or sneezing. And what about basal, non-emotional tears, which are always present and lubricate and protect the eye? A next step in this work is to establish the behavioral and physiological parameters of the tear response in both men and women, and ultimately to identify the chemical stimulus and its sensory receptor.

While waiting on the results, it may not be premature to consider a point that men have long suspected; tear-jerkers are not ideal date movies.

~

Tear research prompted memories of my scientific youth, when I worked in the lab of my mentor, Rita Levi-Montalcini, discoverer of nerve growth factor (NGF). NGF is a naturally occurring chemical agent necessary for the development and survival of neurons.[12] The largest concentration of NGF is in the salivary gland, which produces saliva. Could the tear-producing lacrimal gland also contain NGF? A literature search revealed that NGF is indeed present in the lacrimal gland and its secretion, the complex chemical soup of tears. The story gets better.

Several lines of evidence suggest that the NGF in tears has medicinal functions.[12] The NGF concentration in tears, cornea, and lacrimal glands increases after corneal wounding, suggesting that NGF plays a part in healing. More directly, the topical application of NGF promotes the healing of corneal ulcers[13] and may increase tear production in dry eye.[14] It follows that a deficit of NGF-bearing tears, whether due to the inadequacy of synthesis, release, and/or utilization, may cause ocular pathology. (No one has contrasted the NGF composition of basal and emotional tears.) Although more of a scientific long shot, I suggest that tears bearing NGF have an antidepressive effect that may modulate as well as signal mood.[15]

Non-emotional, healing tears may have originally signaled trauma to the eyes, eliciting caregiving by tribe members or inhibiting physical aggression by adversaries.[12] This primal signal may have later evolved through ritualization to become a sign of emotional as well as physical distress. In this evolutionary scenario, the visual and possibly chemical signals of emotional tears may be secondary consequences of lacrimal secretions that originally evolved in the service of ocular maintenance and healing.

Emotional tearing is a uniquely human and relatively modern evolutionary innovation that may have left fresh biological tracks of its genesis.[12] The contrast of the human lacrimal system with that of our tearless primate relatives may reveal a path to emotional tearfulness that involves NGF. NGF may be both a healing agent found in tears and a neurotrophin that plays a central role in shaping the neurologic circuitry essential for emotional tearing during development and evolution. A lesson of NGF research is that pursuit of the scientific trail can lead to serendipitous discoveries both broad and deep.[16] Emotional tears may provide an exciting new chapter in the NGF saga, and vice versa.

The tear effect has implications beyond emotional tearing and vocal crying. Efforts to read the face have existed from antiquity and were a subject of Charles Darwin's 1872 classic *Expression of Emotions in Man and Animals*. The quest continued as a significant theme of twentieth-century psychology, with Paul Ekman as a leading investigator. On the basis of decades of laboratory and cross-cultural studies, Ekman proposed a group of six *primary emotions* recognized by people of all races and cultures—happiness, anger, surprise, fear, disgust, and sadness—and developed detailed measures to describe these emotions and separate the "true" (spontaneous) from the "false" (faked).[17] The details of this and competing stories are still being worked out by the energetic and sometimes contentious students of facial expression, but the chapter about tears is largely unwritten. The tear effect demonstrates that the traditionally considered neuromuscular mechanism of facial behavior is limited in its ability to portray the full range of emotional nuance. The investigation of visually observed tears has earned a place on the research agenda. Much of the research about the facial communication of emotion merits replication with tears. *Tears change everything.*[5]

Before abandoning my role as provocateur, I offer some thinking points. Given that emotional tears are uniquely human and recently evolved, we may be witnessing an evolutionary process now in progress, when the intermediate steps are still visible and before the loose ends are tidied up. This may explain tearing in such disparate acts as crying and sneezing. Check back in a hundred thousand years to see how the innovation of emotional tears is working out.

The enigmatic presence of tears in sobbing, yawning, laughing, anguish or pain, and sneezing may reflect more than a superficial relationship, neurological anomaly, or emotional overflow. All may share unappreciated neurological kinship as emotions, near-emotions, or proto-emotions. Is the yawn a facial expression of the emotion of boredom or sleepiness? Is the laughing face an expression of happiness or exuberance? And what about the sneeze, which in many ways resembles the yawn or sexual climax? If my proposal is discomfiting (it's meant to be), consider that the criteria for "emotion" are a bit fuzzy, established by vote, and sanctified by tradition. Consider further that all tearful acts are honest signals that are hard to fake, like the "true" Duchenne smiles noted by emotion researchers;[18] most people can't fake laughs, yawns, sneezes, anguish or pain, or tearful sobs. Breakthroughs in understanding the evolution of facial expression may come from such behaviors that are beyond the traditional fold. They force rethinking of issues that are often rooted more in folklore than nature.

~

Facial signals, we have learned, are often ambiguous, and we need all available cues to make sense of them, including tears. I conclude with an exploration of emotional expression in the arts and the use of the arts to inform science. Contrary to popular opinion, art is a very practical, empirical affair, with the artist sorting through the kit of available techniques to find those that can get the job done. There may be more bad science than bad art, but scientists can finesse a problem with florid prose and earnest factness; artists, on the other hand, must deliver convincing goods to a satisfied, paying customer.

However skilled they may be, painters face a daunting task in trying to portray on a canvas the spectrum of human emotion, from joy to despair. How can one accurately re-create a dynamic facial expression, a four-dimensional act of height, width, depth, and time, on a two-dimensional surface having only height and width? A smile, for example, is more than a static posture; it is a complex pattern of fleeting neuromuscular events distributed in space and time.[19] Smiles don't instantaneously spring forth in full bloom, a source of facial whiplash and flying saliva. The artist must settle for an arbitrary snapshot of the smile in mid-flight, a frozen instant that must convey everything. This time sample is often not enough, and the painted face may not convincingly communicate the desired facial expression or emotion. Artists appreciate this challenge and may avoid the risks of unintentionally cheesy, artificial, or comic portrayals by using muted expressions and looking beyond the face for emotional nuance and power.

Context provides what the face alone cannot. Crucifixion scenes in religious art, for example, cue the grieving at

Golgotha; depictions of wedding parties cue the gaiety of the guests. Removed from context, the faces in such paintings may lose emotional impact and valence. Does a visage portray anguish, existential despair, or merely intestinal distress? The earlier chapter about yawning demonstrated that an image of the isolated gaping mouth of a yawning person is an ineffective stimulus for contagious yawns, because it could be interpreted as showing yelling, singing, sneezing, or sexual climax instead of yawning. Thus, yawns are hard to paint. Body posture is another cue of emotion, whether the stooped form of sorrow, the taut body primed for battle, or the outstretched limbs of a person in full yawn. The prop of a handkerchief or hand to the face suggests sadness and tears. Such *simultaneous context cues* can all be viewed at once in a single frame.

All representational painting involves a more pervasive illusion, tricking the viewer into believing that a two-dimensional surface possesses the third dimension of depth. The discovery by painters of what the eye sees and how to realistically convey this information with brush or pen is one of history's great achievements, culminating in the triumphs of the Dutch and Flemish Old Masters during the mid-1500s. But even these masters of two dimensions could not conjure the fourth dimension of time that is part of a typical emotional display.

Unlike the time-locked spatial art of painting, theatrical performance and cinema exploit time to display emotion in the three dimensions of the stage or the two dimensions of the flat cinema screen. They can show the dynamic act of smiling, not just a snapshot of a smile at its zenith. In addition to the simultaneous context cues in painting, stage and cinema scenes provide *temporal context*, information about events that come before and after. While viewing a film of a Sherlock Holmes mystery, for example, you are unlikely to encounter the Busby Berkeley dancers, but an appearance by Professor

Moriarty is not out of the question. Going for a swim has very different expectations in *From Here to Eternity* than in *Jaws*.

Lev Kuleshov, theorist and pioneer of early Soviet-era montage cinema, provided a demonstration of temporal context known as the "Kuleshov effect." Although originally offered as a demonstration of the power of cinematic editing, it's relevant both for theater and for making sense of everyday human behavior. Kuleshov presented a film that alternated shots of the expressionless face of tsarist movie star Ivan Mosjoukine with shots of a bowl of soup, a girl, and a child in a coffin (accounts vary). Identical footage of Mosjoukine's face was variously interpreted as a subtle but masterful expression of hunger, desire, or grief, depending on what he was presumed to be witnessing. A search for "Kuleshov effect" on YouTube provides many variants, including what may be parts of Kuleshov's original in the Spanish-language documentary series *Amar el Cine*. Of note is director Alfred Hitchcock discussing editing techniques and his own Kuleshov demonstration in a clip titled "Hitchcock Loves Bikinis." The YouTube clips are of mixed quality but well worth a look.

Film actors benefit from extreme close-ups that capture misty eyes, quivering lips, and other nuances that would be lost on the theater stage. Silent film actors, lacking voice, relied on necessarily bold face expressions, gestures, and movements. Imagine the challenge of silent screen actors trying to adapt to the talkies, the theme of the great Hollywood musical *Singin' in the Rain*. A challenge to stage actors is that, unlike their counterparts in film, the distant audience does not have a good view of their face; tears shed for their art may be for naught. The burden of emoting onstage is carried to a greater extent by posture, voice, and gesture, such as bringing the hand to the face to dab a nonexistent tear, a body racked by sobs, or moving in a slow, hesitant manner.

Painting, theater, and cinema all struggle to convey emotion within the limits of their medium, whether two, three, or four dimensions. If two-dimensional painting seems restrictive, consider that the completely face-free auditory media of radio drama ("theater of the mind") and music convey emotion using the single dimension of time.[20]

~

My musings about crying, tears, and faces are summarized in an observation about the human predicament.

Daily, we are actor and audience in our own unending, life-and-death drama of unknown plot. We are left guessing our way through life, grasping at fleeting social and cognitive cues that we can barely appreciate and never master, trying to read inscrutable faces, known and unknown. In this uncertain environment, a tear changes everything.

5

whites of the eyes

"Don't shoot until you see the whites of their eyes," said General Israel Putnam to his American militia when facing British forces at Bunker Hill on June 17, 1775. This military anecdote is the best-known reference to the sclera, the eyeball's tough, white outer coating. The sclera deserves better. Like the emotional tears of the previous chapter, it's a uniquely human medium that transforms social signaling. The human sclera signals emotion, health, age, disease, and gaze direction, cues unavailable to our dark-eyed primate cousins. Our white sclera is also the reason why eyedrops that "get the red out" are really beauty aids.

Considering the sensory organ of the eye as an organ of communication requires some inferential leaps. I will prepare the way by sharing my own meandering path of discovery.

In graduate school, I had the good fortune to collaborate on some research projects with Jay Enoch, then professor in the Department of Ophthalmology of the Washington University School of Medicine.[1,2] Jay—a gentleman, visual scientist, and clinician of the first rank—is memorialized here as "Old Iron Eyes" for reasons that will become apparent. Our research required participants to wear thick, powerful contact lenses, and it seemed more efficient for us to manually apply and remove the contacts than for them to learn this skill; stressed, teary-eyed participants would not be good observers. Readers wearing contacts can empathize with their own first encounters with these lenses—it's not normal to put a large foreign object in one's eye, and copious tears are reflexively released to flush it out.

Jay suggested that I practice lens application and removal on him and I applied and removed the lenses every minute or so until I eventually mastered the task, many trials later. This made me a bit queasy, and I teared up a bit, but Jay seemed unconcerned, once chatting away on the telephone while I leaned over him, applying and removing the contacts. Let's face it, touching another person's eyeball is a special kind of intimacy. His nonchalance reduced but did not eliminate my apprehension. However, another year in this department might have done the trick: students could earn extra cash by harvesting cadaver eyeballs, which were stored in glass jars beside our lunch sacks in the lab refrigerator.

Working with Jay and my hardy experimental subjects alerted me to the sympathetic tearing triggered by observing the discomfort, eye redness, and tearing of other people. My next revelation came years later, when I was a professor at my present university.

A soon-to-graduate academic advisee dropped by my campus office to announce her acceptance of a dream job as a sales representative for a major drug company. Jennifer was, as usual, cheery and upbeat, but she had a distressing case of pinkeye (conjunctivitis), a viral eye infection that made her eyes watery and inflamed, and she dabbed tears from her cheeks with a tissue. I tried to avert my gaze, wishing that she had stayed at home. My own eyes immediately started to burn and water, and I worried that I might catch her highly infectious virus. Slowly I realized that I was experiencing an unrecognized form of empathy involving eye redness and tears, and I made a mental note to pursue this effect someday. A related phenomenon was suggested by my wife, Helen, a professional musician and fellow observer of the human condition. Helen suggested that I examine over-the-counter eyedrops and their advertising.

Are you straining to read this small print? Need better light? Get too little sleep? Allergies acting up? Nonprescription eyedrops may be just what you need for those tired, burning eyes. Ad copy promises "fast relief of redness of the eye due to minor irritation . . . caused by conditions such as smoke, dust, other airborne pollutants, and swimming." The active ingredients in eyedrops really do "get the red out" by shrinking the eye's superficial blood vessels, primarily those of the conjunctiva, the thin, transparent membrane covering the sclera. Eyedrops use the same vasoconstrictor agents (tetrahydrozoline hydrochloride or oxymetazoline hydrochloride) put in nasal sprays to dilate airways and improve breathing. As when these substances are used as nasal decongestants, tolerance develops quickly, and a rebound effect is experienced when they are discontinued. But there is more to the eyedrop story than getting the red out.

Consider your experience with television and magazine ads for eyedrops. You are eyeball to eyeball with an actor with teary, inflamed eyes. Not a pretty sight. Does mere exposure to the ad make your own eyes feel a bit tired? Burning? Itchy? Watery? Do you crave a product that "soothes," "cools," and "refreshes"? Would you have experienced such ocular discomfort without viewing the ad? Or did the ad prompt self-monitoring and subjective symptoms that require relief?

Does even reading about this topic cause discomfort? And what about the image of eyedrop application, a common feature of eyedrop ads that is itself a stimulus for tearing? Do these ads trigger recall of past application of drops in your own eyes, the associated stinging and discomfort, and subsequent relief?

Eyedrop ads are insidious, tapping a response that I noticed years before with Old Iron Eyes and my student with pinkeye. The power and immediacy of the response suggest that more is involved than a mindful, cognitively mediated sharing of distress. You don't decide to tear up in sympathy—the urge to do so happens spontaneously and may have an innate basis. Tears cause tears. Red eyes cause red eyes.

It makes perfect biological sense that we have evolved a broadly tuned and easily activated defense mechanism to protect our eyes, especially the cornea, the eye's exposed, highly sensitive, and fragile first lens. The smallest stimulus, physical or psychological, may trigger self-monitoring and a copious bath of lubricating, soothing, NGF-containing, antibiotic tears.

To experience psychological stimuli for ocular self-monitoring firsthand, search Google or Flickr for "eyedrops," various brand names of eyedrops, "pinkeye," "conjunctivitis," and the like. You will encounter some unsettling but compelling fare.

When the Whites of the Eyes Are Red

After decades of intellectual incubation, it was time to cease collecting anecdotes about the impact of scleral coloration on behavior and put these ideas to the test in the lab.[3] My research team evaluated eye redness as a visual cue by contrasting the ratings for "sadness," "healthiness," and "attractiveness" of one hundred eye images with normal whites against one hundred copies of those same eyes whose sclera were reddened by digital editing. Sadness, healthiness, and attractiveness were selected as a representative, but not exclusive, sample of emotional, biological, and social dimensions of the observed eye.

Individuals with reddened sclera did indeed appear sadder, less healthy, and less attractive than those with whiter, untinted sclera. Red eyes join facial expression and emotional tears as visual facial signals of sadness, with cues of health and attractiveness a bonus. However, context may be necessary to choose between these options and discern the cause of redness. When encountering a red-eyed friend, it may be unclear whether you should offer sympathy or medical assistance. Red eyes, for example, may be the result of grief, allergies, irritation, or infectious disease.[4]

Red, bloodshot eyes are the most common ocular complaint to physicians,[4,5] who have understood the importance of scleral color in physical examinations from at least as early as the Hippocratic era, several centuries BCE.[6] In varying degrees, scleral redness is a symptom of numerous pathologies of the eye and body, both major and minor, acute and chronic, including conjunctivitis (viral and bacterial),[4,6,7] chemical and physical irritation of the eye, dry eye, eye trauma, episcleritis, glaucoma, allergy,[8] hypertension, diabetes,[9] sickle cell disease,[10] autoimmune disease, rheumatoid arthritis, sleep deprivation, weeping, and drug use.[11,12]

Our most recent experiment added yellow to the palette of evaluated scleral tints. Yellow is clinically interesting because it signals liver disease (jaundice)[13] and aging.[14,15] Eye images with digitally yellowed sclera are rated as less healthy, less attractive, and older than identical control images with untinted sclera.

Beauty Is in the Eye of the Beheld

Beauty is said to be in the eye of the beholder, but here we discover the reverse: attractiveness is in the eye of the beheld. Do beautiful people ever have unattractive eyes? The window of the soul is uninviting if clouded by cataracts or infection. Clear, bright eyes join thick, shiny hair, glowing, blemish-free skin, symmetry, and such sexually dimorphic traits as (for women) large eyes, full lips, high cheekbones, and small chins, and (for men) small eyes, thin lips, strong chins, and prominent brows as universal, hard-to-fake signs of health and beauty.[16–22] Primatologist Michael Tomasello and colleagues note in passing, "One can easily imagine that white sclera signal good health and therefore good mates. However, to our knowledge, there is no evidence for this hypothesis."[23] The present research provides the necessary evidence.

Beauty is a practical affair. Standards of beauty vary across cultures and through history, but youth and healthiness are almost always in fashion because they are associated with reproductive fitness.[17,21,24–25] Inflamed, teary eyes are prominently posted advertisements of disease, allergy, irritation, and drug use. Want to spend time around people with such symptoms, let alone mate with them?

Eyedrops are a beauty aid that whitens the whites by shrinking superficial blood vessels of the conjunctiva.[26] There is no comparable demand for a vasodilator to produce flamboyant crimson sclera. Vasoconstrictive eyedrops have a short-term cosmetic effect that hides underlying physiological causes, from sleep deprivation to marijuana use, but their effect is self-limiting because of tolerance and rebound. Eyedrops join the ancient drug belladonna (the word means "beautiful lady") as an ocular cosmetic. Belladonna enhances beauty by dilating the pupil, mimicking a person's response to interesting and sexually arousing stimuli.[27]

What cannot be changed pharmacologically can be masked behind dark eyeglasses, another product that benefits from the sclera's honest signals of emotion and health. Wearing dark glasses indoors is often considered rude because it erects a social barrier. Jazz musicians who made wearing dark glasses in dark environments fashionable sought a barrier of a different kind—one that hid the pinhole pupils of a heroin addict or the bloodshot eyes of a marijuana user.[11] Red or atypical eyes trigger a response in those who view them. Unsolicited comments from a few of our subjects suggest that images of red eyes are both unsightly and may produce discomfort, ranging from increased monitoring of their own eyes to a hint of sympathetic tearing. Some blind individuals avoid possible negative audience responses to their eyes by wearing dark eyeglasses.

In a scleral health and beauty contest, plain white trumps red or yellow, but the potency of a white sclera has limits. Eye images with super-white sclera produced by digital editing were rated as no more healthy or attractive than control images of normal, untreated eyes. A ceiling effect may be involved with our fairly young, healthy sample: the normal white sclera may already approach the maximum amount of ocular health and beauty. However, if preparing an important portrait, a little tasteful Photoshop work to brighten the eyes will do no harm. People with dark irises and skin such as Africans[28] and Inuits[29] are most likely to benefit because they are much more likely to have obvious scleral spots than those with blue irises and light skin. For in-person appearances, the eyes can also be brightened cosmetically by using dark mascara to enhance the visual contrast with the white sclera, although some males may prefer to stick with eyedrops.

Evolution of White Sclera and Signaling

"Get the red out," the mantra of over-the-counter eyedrops, only makes sense if you can see the red, which requires the white ground of the human sclera, the region of the eye surrounding the pigmented iris. I was excited to discover Hiromi Kobayashi and Shiro Kohshima's exhaustive comparative analysis,[30] indicating that humans are the only primates with white sclera. The brown sclera of our primate cousins masks redness produced by dilation of their eye's conjunctiva or other superficial blood vessels, and diminishes associated cues of emotional and physiological state. As with emotional tears, discussed in the previous chapter, a white sclera provides a uniquely human medium for signaling.

The evolution of our prominent whites also permits the signaling of gaze direction, with the white sclera framing the darker iris and pupil and increasing the visibility of orbital movement.[23,30–32] Human eyes do not require a shift of head or manual pointing to communicate the position of an interesting object, event, or person. A glance may be sufficient. The object of another person's gaze may even be you, the recipient of signaled interest, affection, distrust, or aggression. The dark sclera of nonhuman primates diminishes gaze direction signaled by ocular movement, but this may not always be a disadvantage. The dark sclera may be an adaptation to mask intention and gaze direction.

The stimulus potency of all aspects of the human face makes deception hard, at or away from the poker table. We have neurons that respond selectively to stimuli of eyes and faces.[32,33] These visual processes may provide an unconscious, automatic sensation of being watched, and encourage altruism.[34–36] The potency of gaze is sufficient that even the video eye images of a robot—Cynthia Breazeal's celebrity

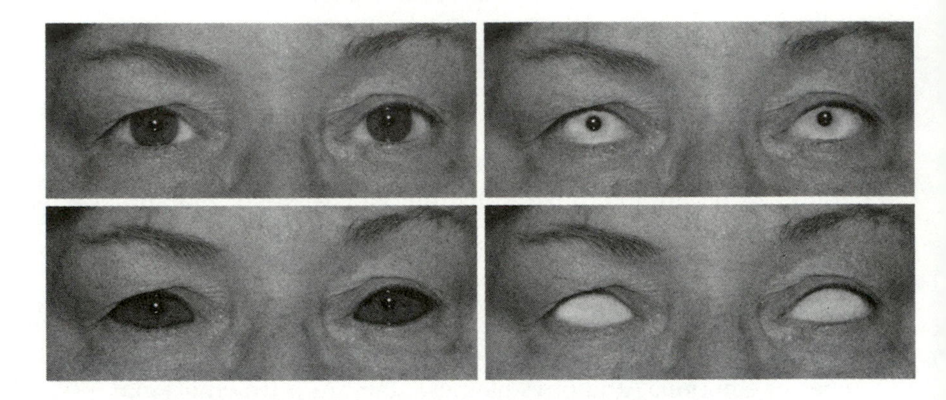

figure 5.1 *Digitally edited images demonstrate the visual impact of the uniquely white human sclera (the white of the eyes). Contrast normal eyes* (upper left) *with those having a dark, ape-like sclera* (lower left), *sclera extending to the pupil* (upper right), *and sclera totally occluding the iris and pupil* (lower right). *Abnormal sclera appear startling and often disturbing.*

robot Kismet—prompted charitable behavior by anonymous individuals.[37]

The impact of white sclera on eye-related visual cues is demonstrated by digital manipulation of a human eye image (Fig. 5.1). When the sclera are darkened, in this case by expanding the hue of the iris into the normally white sclera, the now dark, primate-like sclera become deficient in signaling gaze direction and redness. Even the pupils are difficult to detect. Next is a bizarre variant in which the white sclera are extended to the edge of the pupils, leaving them black dots afloat in a milky sea. Whiting out the entire orbit produces an apparently sightless eye, as if the irises and pupils are masked by a massive cataract. Lab members found these eye variants disturbing. I took the opposite position—grotesque can be good when, as in these demonstrations, it signals an anomaly in a critical and familiar perceptual cue. The trivial and ordinary would go unnoticed.

As a communication medium, shifts in scleral redness are limited by the unmeasured but probably slow rate of conjunctival capillary dilation and contraction, processes better suited to cueing tonic physiological and emotional state (e.g., sadness, healthiness) than fleeting, phasic events (e.g., flashes of anger, fear, arousal). The response may be more sluggish than the dilation and constriction of the iris that signals ambient light level, stimulus interest, and sexual arousal. The human sclera is certainly not capable of the almost immediate, spectacular chromatic displays of the cuttlefish and its cephalopod relatives, displays that these sea creatures use for camouflage and communication.

The visual cues considered in this and the previous chapter—scleral color, gaze direction, pupil dilation, emotional tears—indicate that the sensory organ of the eye has acquired a secondary role as an organ of communication. The richness of perceptual cues associated with the eyes indicates why they are a focus of visual attention.

6

coughing

A cough provides a pneumatic blast that clears the throat and lungs of irritants and debris, an act necessary for good health, or, in emergencies, even survival. When chronic, coughing is the leading medical complaint, usually of upper respiratory illness. Despite these impressive physiological and medical credentials, coughing seems an unlikely topic for a scientific detective story. Where is the passion to understand airway maintenance? However, several factors recommend it. Coughing is a simple behavior that can be produced on command and is easy to measure. Easy is good. No waiting around for rare, unpredictable behaviors such as sneezing or hiccupping. Better yet, with expectations so low, something of interest is bound to show up. We will start by describing coughing.

The cough involves an initial deep breath, followed by an exhalation driven by contraction of the abdominal muscles and diaphragm. Thoracic pressure rises because the exhaled air is dammed against the closed glottis. Sudden opening of the glottis produces an explosive release of the trapped,

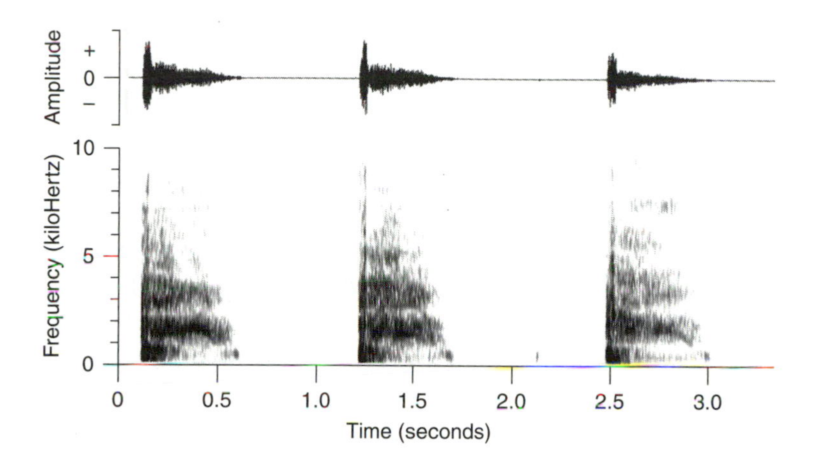

figure 6.1 *The waveforms* (upper) *and frequency spectra* (lower) *of three voluntary coughs of an adult male. Note the stereotypy, noisy sound, and behavioral features of the cough, especially the explosive onset associated with glottal opening at the leading* (left) *edge of each cough. Coughs lack the strong harmonic structure of laughs, another explosive vocalization (see Fig. 2.2). The three coughs are placed on the same baseline for purposes of comparison; the intervals between coughs do not reflect actual intervals between coughs.*

pressurized air. The process is like the inflation and bursting of a balloon. Without the buildup in air pressure against the closed glottis, a cough would be little more than an unproductive sigh.

The sound and some behavioral properties of three voluntary coughs by an adult male are summarized in Fig. 6.1. The explosive exhalation associated with glottal opening is prominent in the leading edge of the waveforms (upper) and sharp, high-frequency onset of the spectra (lower). These coughs are stereotyped in form and have a duration of about one-half second.

The sound of coughing, like the sound of an ailing automobile engine, is a means of diagnosis—croupy (subglottal swelling), brassy (aneurysm of the aortic arch), dry or productive, single cough or in clusters (paroxysmal)[1]—but this interpretive art is being displaced by less subjective modern medical technology.

The blast of air during a cough or sneeze may approach the wind speed of hurricanes or tornadoes, although its exact velocity is neither agreed upon nor easily measured. The more leisurely breeze of normal respiration is insufficient for clearing the airways. As vocalists know, coughing impairs the voice and should be avoided. Violent coughing can cause headaches and, in extreme cases, even crack ribs, especially in older individuals with osteoporosis. Coughing has many causes. In a sample of 102 patients between the ages of three weeks and fifty years who were coughing continuously, the leading causes of coughing were postnasal drip (41 percent), asthma (24 percent), gastroesophageal reflux (21 percent), chronic bronchitis (5 percent), bronchiectasis (4 percent), and miscellaneous (5 percent).[2]

Coughs can be reflexive or produced on demand. Voluntary control is important. If you were gagging, would you like to wait for an uncertain reflex to clear a food particle from your throat? The voluntary control of coughing is indicated by its relatively short reaction time (1.7 seconds), as contrasted with the very long reaction time (8.1 seconds) for the usually vain attempt to sneeze on command. When asked to cough, 95 percent (98 of 103) of study participants were able to do so within ten seconds. The ease of voluntary coughing is reflected in its position toward the easy-to-play right side of the Behavioral Keyboard in the appendix. Humans may be the animal kingdom's leading coughers, a possible result of the respiratory demand of our bipedality and associated upright posture, and our ability to cough both reflexively and voluntarily.

Cough Variants

The voluntary control of coughs makes them easy to study. We don't have to stimulate them via irritants or wait for them to occur spontaneously. You can just cough. I offer several cough variants that demonstrate basic properties of coughs and provide contrasts with airway maneuvers in other chapters (yawning, sneezing).

Mouth coughs are produced by pinching your nostrils closed and expelling air through your mouth.

Clenched-teeth coughs are produced by clamping your jaws shut and expelling air through your teeth.

Nose coughs are produced by sealing your lips and expelling air through your nose.

Open eye coughs are produced by propping your eyes open with your fingers and expelling air through your mouth as in a normal cough.

All four of these cough variants are relatively easy to perform, reflecting a flexibility lacking when similar interventions are tried with yawning (Chapter 1) and sneezing (Chapter 7). Coughing (voluntary and reflexive) may have evolved primarily to service the bronchia and throat, complementing sneezing (involuntary), which is more dedicated to clearing the nasal airways. The mouth cough approximates the everyday cough. The nose cough (or snort) is often used by some individuals to launch "snot rockets," an effective if unattractive and unsanitary means of nasal clearance. It's best left to adepts.

Cough CPR

Coughing has consequences beyond clearing the airways. The violence of coughing compresses the heart, making possible "cough CPR," a potentially lifesaving maneuver for victims experiencing ventricular fibrillation and perhaps other cardiac emergencies.[3] (CPR is cardiopulmonary resuscitation, a widely used emergency procedure involving repetitive manual pressure applied to the closed chest of the victim.) J. Michael Criley and colleagues at Harbor General Hospital in Torrance, California, reported that three of eight of their patients who experienced ventricular fibrillation while undergoing a coronary catheterization procedure were able to remain conscious and alert for twenty-four to thirty-nine seconds after the onset of fibrillation by coughing every one to three seconds. This was sufficient time to administer cardiac defibrillation. Remarkably, the average systolic arterial blood pressure of their coughing patients was more than twice that produced by traditional cardiopulmonary resuscitation, indicating that the procedure was highly effective in producing blood flow after circulatory arrest. It's helpful that most cough-induced blood flow is headward (cephalad). The maintenance of consciousness by the coughing patients indicated that their brains received adequate blood flow. This is a critical and promising sign because the brain is the first organ to succumb to the effects of cardiac arrest and associated hypoxia.

Cough CPR may be useful to victims trained to recognize the early stages of ventricular defibrillation and begin coughing while still conscious. This maneuver could delay cerebral hypoxia and unconsciousness, provide time to seek help, and earn vital seconds of cerebral perfusion that make the difference between survival with normal brain function, survival with hypoxic brain damage, and death. Cough CPR has advantages over traditional CPR, which involves chest compressions; it is self-administered (self-CPR), both cardio and pulmonary resuscitation are delivered because a breath is taken between each cough, and it avoids the rib and sternal fractures and other damage sometimes produced by manual chest compressions.

Now the controversy and warning. On its website, the American Heart Association "does not endorse 'cough CPR'" and stresses that it "should not be taught in lay-rescuer CPR courses because it is generally not useful in the prehospital setting," noting that the "usefulness of 'cough CPR' is generally limited to monitored patients with a witnessed arrest in the hospital setting."[4] The AHA is obviously concerned that attention to the specialized utility of cough CPR will deflect attention from critical procedures such as seeking immediate help via 911 and taking aspirin. Some websites even suggest that cough CPR is a hoax. However, three decades after his initial report, J. Michael Criley still supports the cough CPR procedure,[5] noting that it can occur outside a monitored setting, can be taught to high-risk patients, and can save lives.[6]

The motive for considering cough CPR here is not to engage a debate best left to cardiologists but to emphasize the unanticipated consequences of the apparently mundane act of coughing.

Coughing and the Cerebrospinal System

The sonic blast of coughing has consequences that extend beyond the cardiovascular system. Coughing produces a pressure wave that pumps the cerebrospinal fluid (CSF), which bathes our brain and spinal cord, and it may influence blood flow (as noted above) as well as neurotransmitter and hormone release, uptake, and migration.[7] More subtle pulsations of CSF are produced by the heartbeat.[8] Because fluids can't be compressed, pressure applied at one point of a fluid-filed cavity is expressed in another, the principle behind hydraulic systems.

In some individuals, a cough causes a rapid rise in CSF pressure that produces a concussion-like loss of consciousness (syncope). Even a single cough can cause a loss of consciousness within a few seconds, too fast to be the result of ischemic changes associated with blood being squeezed from the brain. Thus the suggestion that the syndrome be termed the more descriptive "cerebral concussion" instead of the traditional "cough syncope."[9]

The hydraulic brain message produced by coughing, sneezing, yawning, and other acts may produce unappreciated secondary behavioral consequences, including alterations of attention, mood, or state of arousal. The coughs that we produce for no obvious reason may not be meaningless displacement activity or tics but an unconscious effort to change our consciousness and behavior in yet unknown ways. Chasing down the consequences of coughing is the kind of reasonable but nontraditional project that falls between disciplines and gets little attention but may provide a good scientific and medical payoff. The competition will be minimal. What's the likelihood, for example, of a psychologist studying coughing, except to learn how to quell presumed psychogenic coughs? The cerebrospinal fluid system involved in this story certainly deserves attention for more than being a source of unwelcome symptoms and samples associated with increased intracranial pressure, debris from infection, strokes, and other brain pathology. When a physician seeks a sample of your cerebrospinal fluid, your life is not going well.

Conversational Coughing

So, when do you cough? Coughs do not burst into a behavioral void but have linguistic and social contexts. Do you cough randomly during the stream of speech, or at linguistically significant moments? And is your coughing influenced by the behavior of others? These considerations take us into uncharted territory and provide fundamental insights into the mechanisms of breathing, coughing, speaking, and socializing.

Anecdotal observations of hundreds of my college students over the past decade suggest that they punctuate their speech with coughs, seldom coughing in mid-phrase. Coughing during speech is grammatically organized, not random. As previously discussed for laughter (Chapter 2), most coughing occurs before and after statements, questions and phrases, the places where punctuations would be placed in transcripts of conversations. I also suspect that the coughing of one person can punctuate the speech of another, although this is probably a weaker effect. Of course, coughing itself can be a vocal signal, as when used to attract attention or to take exception to an annoying comment.

For those doubting the existence of punctuation phenomena, consider that breathing itself punctuates speech—we stop breathing while talking, and take breaths at linguistically significant points.[10] Breathing is grammatical! Punctuation, whether by coughing, breathing, or laughing, indicates the dominance of recently evolved speech over these more ancient respiratory and vocal acts. It is a good thing that punctuation occurs at a low level of awareness. Imagine how complicated your life would be if you had to remember when to cough or breathe. I have not studied the grammatical context of sneezing, but unlike coughing, it may not be subservient to speech. Sneezes overwhelm us. When we sense an approaching sneeze, we hunker down and wait for it to pass.

Musical Coughing

As a lapsed musician, I wondered if the nearly ubiquitous punctuation effect extended to musical phrases. Enlightenment came in the form of an anecdote my wife, Helen, supplied about Diane the coughing pianist. Helen is a piano teacher, so she is exposed to many aspiring musicians, good and indifferent. (As I type this, I hear a student pianist tinkling on the Steinway in our home studio.) Her student Diane is both a good pianist and a prolific cougher, an informative combination. My interest is neither in Diane's very capable Bach nor in her croupy cough, but in *when* she coughs while playing. Diane coughs before and after playing, but seldom in the midst of a musical thought, a different pattern than is typical of beginning students, who cough at any time, indifferent to the integrity of the musical phrase. Before dismissing Diane's behavior as reflecting her discipline to wait until the end of a musical thought or piece to cough, consider the previously noted punctuation effects. With expert musicians such as Diane, music has acquired language-like properties undeveloped in novices, whose pianism is treated as a lower-level motor skill.

Consider, also, coughing patterns of the audience during concert hall performances. Most coughing clusters before and after pieces and movements and so does not disrupt the musical interlude. As found for laughter (Chapter 2), punctuation is present for both performer ("speaker") and audience.

Medical Coughing

The use of coughing in registering a medical complaint is detailed by Julia Bailey, a physician/researcher at University College London interested in doctor-patient communication.[11] Her discourse analysis was based on thirty-three doctor-patient consultations. Coughing is of particular interest because it is both a symptom and a paralinguistic device used by patients to persuade the physician of the significance and doctorability of their complaint, without seeming hypochondriacal or losing face if the cough is deemed a medically trivial result of a cold or other upper respiratory illness. Coughing, as expected, is coordinated with speech instead of occurring randomly, seldom completely disrupting conversation of either patient or physician. The same is true of "pain talk."[12] Patients' cries of pain are placed such that they don't disrupt a clinical consultation. Coughing is sometimes associated with acceptance or resistance of the physician's diagnosis. When told that his chest is clear, for example, a patient may cough as a face-saving, nonverbal response to the challenging news. However, as noted in the introduction, the sound of coughing makes its own significant contribution to medical diagnosis, apart from talking about illness.[1]

Social Coughing

Is coughing as socially disengaged as the knee-jerk reflex? Tap your lifelong experience and recall coughs you have heard at the cinema, during concerts, at lectures, or during exams. Did you hear an isolated cough, or bouts of several? Was it random? Is coughing contagious in the manner of yawning or laughing? Answers to these questions have implications for the sociality of the cough.

Our expert guide to social coughing is James W. Pennebaker of the University of Texas, who has made a career of investigating interesting problems in creative ways.[13] When at the University of Virginia, with the aid of research assistants, he established the baseline cough rate of undergraduate students during lectures. About 29 percent of students coughed at least once during a lecture, with seasonal variations accounting for almost three times as many coughs in February than in April, and large classes having more coughs than small ones. There was also an almost perfect inverse relation between teacher evaluations, course interest, and the amount of student coughing, with the most interesting courses with the best-rated professors having the fewest coughs. It's not a good omen if you are teaching a small class in April to a bunch of coughing students.

Large classes may have more coughs than small ones because there are more coughs to hear, and there is less social inhibition associated with the anonymity of a larger crowd. Coughs of different students tended to cluster, evidence of a social coupling process. Cough clustering was present even when exams were being taken and it was less likely that a common stimulus such as the lecturer or film was being modeled. Proximity was also a factor; the closer a student sat to a cougher, the more likely that he or she too would cough. A low-level coupling process seems to be involved because students had little awareness of coughing, whether their own or that of others, when questioned about events during the previous lecture. Coughing mindlessly triggers coughing in those who hear it.

Pennebaker was prompted to pursue the questions of low awareness by two parents who reported that their young children coughed contagiously while sleeping. The cough of one child would trigger coughs from other sleeping siblings. Thus motivated, Pennebaker shifted his research venue from college classroom to firehouse bunkroom, where he spent a night recording the coughing of fifteen sleeping firemen. Unfortunately, the dozing firefighters spontaneously coughed only twice during the entire night. To compensate for the low rate of coughing, Pennebaker coughed once every thirty to forty minutes in an effort to stimulate coughs. Of his seven experimental coughs, two were immediately followed by single coughs from one of the sleeping firemen. This small sample necessarily lacks the rigor of the contagious yawning and laughter studies, but it deserves follow-up by a nocturnal researcher with access to people in communal sleeping quarters. Alternatively, you can see if your sleeping partner or roommate responds contagiously to your cough.

It is unclear if Pennebaker's coughs are contagious in the manner of laughing or yawning; they may be a pseudo-contagious consequence of self-monitoring, with perceived coughs focusing the audience's attention on tickling in their own throats that must be relieved by coughing. Informal observations indicate that we don't immediately bark back a response to coughs we hear, as we do with contagious laughs, nor do we feel the grinding inevitability of a pending, contagious yawn. The question remains open. You can ponder these issues when next hearing coughs during a musical performance. Is it the flu or the quality of the Mozart?

7

sneezing

Sneezes are humbling. From the first tickling, burning sensations in the nostrils until its explosive climax, a sneeze hijacks our body and commands our attention. Titillating pre-sneeze sensations may wax and wane for many seconds, but we are already in their grasp. A sneeze is nagging and insistent, intrusive and incorrigible, and can't be willed out of existence. Once the process of sneezing is under way, it goes to completion, as with yawns. We are automata under its control.

The exertions of a sneeze are so great that they can put even the mighty at risk. Baseball slugger Sammy Sosa sneezed himself out of the Chicago Cubs' lineup for a month because of a sprained back. Sosa explains, "It would have been better if I had hit off the wall or we have a fight or something, but this . . . you know what I mean?" (Associated Press, May 17, 2004). San Diego Padres pitcher Mat Latos strained his left side by attempting to hold back a sneeze (*Sporting News*, July 16, 2010). But Sosa and Latos were much luckier than a man who experienced a cervical herniation during a sneeze and died shortly after, the victim of an injury akin to hanging.[1] Sneezes have also caused retinal detachment, loss of consciousness (syncope), stroke, miscarriage, automobile accidents, and much else. By these standards, sneeze-produced incontinence and flatulence are mere embarrassments.

For some, the problem is getting sneezes to stop. If one sneeze is debilitating, imagine hundreds of sneezes per hour. A nine-year-old girl produced 237 sneezes in twenty minutes and was forced to leave school because of the disruption.[2] Her sneezing continued at a lower rate in the less stressful home environment. Stress-modulated sneezing was also reported in a boy of fourteen who sneezed for thirty-three days at a rate up to three to six times per minute.[3] Most such cases of intractable sneezing are presumed to be of psychogenic origin because they have no obvious physical cause and do not respond to anticonvulsive agents.[4] Some individuals with a unilateral infarction of the brain stem (lateral medullary syndrome) want to sneeze but can't.[5] They feel the typical tickling sensation in the nose and inhale deeply, but they do not enjoy the resolution of these preliminaries in a climactic sneeze. Several individuals have contacted me with similar complaints about pending yawns that will not come.

Sneezing is usually triggered by irritants that activate sensory receptors in the nose that pass excitation via the trigeminal nerve to the neurological sneeze center in the brain stem.[6] Once a critical threshold is reached, a second respiratory phase of the sneeze is initiated that plays out in a complex piece of universally recognized, neurologically programmed choreography: tilting back of the head, gaping of the jaws, and deep inhaling, followed by a tilting forward and then eye closing and jaw closing during the climactic, explosive exhalation. As with a cough, the explosive exhalation is caused by an initial closing and then sudden release of the glottis. Without glottal involvement, thoracic pressure could not build up and be suddenly released; both the sneeze and the cough would be more yawn-like, producing only a sigh.

Sneezes are noisy; within the confines of a small car they are ear-ringing, reports my wife. Compared to a grandiose sneeze, a cough seems a modest bark, a quickie that lacks commitment. The sneeze probably blasts thousands of particles and millions of bacterial and viral hitchhikers over distances of many meters, the distance and loft time depending on size and weight of the projectile. The speed of air exhalation during a sneeze is high but debated and hard to measure, with estimates ranging almost tenfold, from 100 km/h up to an unlikely 1,045 km/h, almost 85 percent of the speed of sound. If placing a bet, I'd opt for the middle range of the estimates while keeping an open mind.

Sneezes (like hiccups) are difficult to study because they are infrequent and, unlike the cough, can't be produced at will. The involuntary nature of the sneeze is consistent with its very long latency (8.1 seconds) in our reaction time study and corresponding placement on the hard-to-play extreme left side of the Behavioral Keyboard (in the appendix). Only 22 percent (23 of 103) of participants sneezed on command, and these game attempts may not have been true sneezes.

To produce sneezes, sniffing a nasal irritant is a useful approach, although some people can sneeze in response to the less invasive stimulus of light (see below). As a nasal irritant for my personal experiments, I avoided the cliché of pepper and the pharmacological route of histamine spray (nasal sprays are *anti*histamine), opting instead for tobacco snuff, the more sophisticated choice of seventeenth-century European aristocrats. I chose "high toast," a dry, finely powdered snuff known to be especially "sneezy." I gently sniffed a pinch of snuff into one nostril while holding the other shut, then repeated the process on the other side. The snuff has a nice tobacco aroma, but it produced a mild burning sensation in the nostrils that reliably gave rise to one to three sneezes after a latency of up to tens of seconds. There seemed to be a mild nicotine lift after a few sniffs. Some hard-core sneeze researchers—the sneeze fetishists—favor the less courtly mechanical stimulus of the point of a twisted tissue inserted into the nostril. As a young child I once used a similar approach, inserting a feather into the nose of my peacefully sleeping father; he woke with a start, swatting the air around his face. Well, it seemed like a good idea at the time. Such was my first attempt at sneezing research.

The exhalations of my spontaneous and snuff-induced sneezes were primarily through the mouth, a route inconsistent with its presumed function of nasal clearing. The nasal cavity is, after all, the site of stimulation, whether snuff or the mechanical tickle of an insect intruder. Perhaps the demands of human bipedality brought a shift in function from nasal to more general airway clearing. However, a quadriplegic friend reports the sneeze to be useful in nose clearing.

The sound and some behavioral properties of a sneeze are summarized in the waveform (upper) and frequency spectra (lower) of Fig. 7.1. As suggested by the famous acoustic shorthand of "ah-choo," the sneeze is a two-phase respiratory event, starting with an inhalation and climaxing with the explosive release of compressed air initiated by glottal opening. As noted for the cough, without the buildup of compressed air against the closed glottis, the sneeze would be only an unproductive sigh. Contrast the explosive onset of the cough (Fig. 6.1) with the more gradual onset of the sneeze (Fig. 7.1); coughs lack the longer inspiratory prelude of the sneeze. Unlike the cough, the maximum amplitude of the waveform and highest spectral frequencies of the sneeze occur midway in the behavior. Sneezes for this adult male had a duration of about three-quarters of a second and were stereotyped in

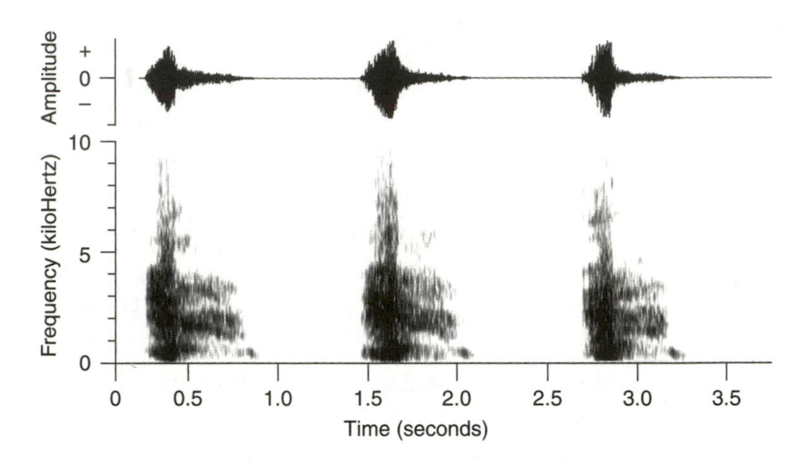

figure 7.1 *Waveforms* (upper) *and frequency spectra* (lower) *of three sneezes of an adult male stimulated to sneeze by sniffing black pepper. Although a powerful airway maneuver, the sneeze lacks the abrupt onset of the cough (see Fig. 6.1), reflecting a tapered inhalation phase before the explosive exhalation triggered by glottal opening at midpoint of the two-stage "ah-choo" of the sneeze. The three sneezes are placed on the same baseline for purposes of comparison; the intervals between sneezes do not reflect actual intervals between sneezes.*

duration and form. Coughs for this same individual were also stereotyped but had a shorter duration, about one-half second.

Variations in Sneezes

I explored the mechanics of sneezing using variants that you can try during your next sneezing episode. The key is remembering what to do when a sneeze is pending. These demonstrations and their results are anecdotal and based on self-observation by me and a few friends, but they seem reliable and may prompt your thinking and own studies.

The *mouth sneeze* is performed by pinching your nostrils closed when you feel a sneeze starting, forcing the preliminary inhalation and terminal exhalation to occur through your mouth. My mouth sneezes felt normal, suggesting that human sneezes play a relatively minor, inessential role in nasal clearing. Sans pinched nostrils, this is probably the typical normal sneeze.

The *clenched-teeth sneeze* is performed by closing your jaws when you feel a sneeze starting, forcing the preliminary inhalation and terminal exhalation to occur through your teeth. My sneezes blasted their way thorough my clenched teeth, but they did not feel normal. (Clenched-teeth yawns, in contrast, are typically blocked in midcourse; see Chapter 1.)

The *nose sneeze* is performed by sealing your lips when you feel a sneeze starting, forcing the preliminary inhalation and terminal exhalation to occur through your nostrils. I could perform nose sneezes, but this variant is discouraged for all but the most committed sneeze sleuths. The nasal passage is narrow, producing greater than normal pressure in the airway during the explosive exhalation, potentially damaging eardrums and other structures. But it certainly clears the nostrils! The nose sneeze is probably the ancestral form. On the path to bipedality, we may have evolved a sneeze less specialized for nasal clearance. However, chemical and mechanical irritation of the nasal passages remains the human sneeze stimulus despite our reduced ability to expel it by sneezing, except when a nose sneeze is performed with closed lips.

The *open-eye sneeze* is performed by propping your eyes open with your fingers when you feel a sneeze starting. In this variant, eye closure, a normal component of the motor program of sneezing, is blocked. (Eye closure is presumably adaptive because it protects the eyes from the expiratory blast.). Folklore suggests that sneezing with open eyes causes the eyes to pop out. When I tried it, my eyeballs remained in place, but I made an unanticipated discovery: pending sneezes were sometimes stopped in their tracks. As a class exercise, I asked students in one of my large lecture classes to examine the effect of sneezing (and yawning) with their eyes propped open, but to avoid biasing their reports I provided no further instructions. Without guidance, several students reported the sneeze- and yawn-blocking effects of keeping the eyes open. This investigation is still under way. What is your experience? Coughs lack the eye-closing component typical of sneezing and are not impeded by holding the eyes open.

The Sneeze as a Fast Yawn

The evolution of any behavior is difficult to track, especially a complicated one such as sneezing. The complex choreography of the sneeze is unlikely to have emerged fully formed; it's probably a modification of an existing motor act. I propose that the sneeze is derived from the yawn, with the *sneeze as a fast yawn*. The origin of sneezing from yawning is not as fanciful as it may seem on first hearing, and the exploration of this proposal contributes to our understanding of both acts. The yawn meets two conditions for being the ancestral form: priority and similarity. The antecedent must evolve first and resemble the derivative act. As considered in Chapter 1, the yawn has great antiquity, as indicated by its near ubiquity among vertebrates, from fish and reptiles to primates. The antiquity of sneezes is harder to establish, but the sneeze probably evolved more recently in terrestrial (non-aquatic) species that would benefit from clearing the nostrils of dried secretions, foreign particles, and invading organisms. Sneezing has been reported in a variety of reptiles, especially iguanas, but not in fish. The sequence of development provides other, indirect evidence of antiquity, with the oldest act emerging first. By this standard, yawning is easily the most ancient, being present by the end of the first trimester of prenatal human development. Sneezing develops later and is obvious in newborns.

Regarding the criterion of similarity, yawns resemble sneezes in many but not all respects. Both are universally recognized human instincts of both sexes. Both involve the playing out of similar long, complex, anatomically distributed (mouth, thorax, eyes, etc.), neurologically orchestrated motor programs. Both begin with a long inspiration, tilting back of the head, and gaping of the jaws, and both climax with eye closing, tilting forward of the head, and an exhalation—the slow sigh of the yawn or the shorter, explosive discharge of the sneeze. The facial expression with gaping mouth during the early inspiratory phase is similar in both acts, but the route of air intake is different. In sneezes, inspiration can be through either the nose or the mouth, in contrast to yawns, where it is almost exclusively oral. (Most people find it impossible to perform a nose yawn.) Neither yawning nor sneezing is under voluntary control, as indicated by the very long reaction times of subjects vainly trying to perform them (yawns = 5.7 seconds, sneezes = 8.1 seconds; see the appendix). In contrast to sneezes, coughs are relatively brief, involve a simpler motor program, and are performed either voluntarily, as indicated by a short reaction time (1.7 seconds), or reflexively. Yawns and sneezes differ in stimuli. Whereas the sneeze has a well-known physical stimulus (nasal irritants), yawns do not; yawns are contagious, and sneezes are not.

Sneezes lack contagiousness, a striking feature of yawns and perhaps even of coughs. Given the similarities in appearance between sneezes and yawns (i.e., tilting of the head back and then forward, gaping of the mouth, closing of the eyes), it's curious that the brain does not produce at least a few erroneous crossover contagious yawns in response to observed sneezes. Sneezes are apparently treated as a separate perceptual category, perhaps due to the more rapid exhalation phase. It's just as well that sneezing is not catching; a chain reaction of contagious sneezes would be spectacular, disruptive, and unsanitary.

Although sneezes and yawns are topographically similar, with speeded-up video images of yawns superficially resembling sneezes and slowed-down sneezes resembling yawns, high-resolution video and electromyographic analyses will be required to ultimately settle this issue. Our preliminary findings indicate that the project will be worth the effort and solve a fascinating problem in human neurobehavioral evolution. The analysis could also include a contrast with orgasms, the facial expression of which resembles the early inspirational stage of both sneezes and yawns. Next time that you see magazine ads for allergy medications, check out the faces of the pre-sneeze (pre-orgasmic?) individuals that are typically featured.

The Curious Case of the Photic Sneeze

About a quarter of the adult population has the *photic sneeze reflex*, the heritable (autosomal dominant) tendency to sneeze in response to bright light.[7] The photic sneeze is usually a one-shot affair, with only the first light exposure stimulating a response; subsequent light exposure over the short term is ineffective. The photic sneeze is usually considered as a behavioral curiosity, prompting occasional scholarly review, letters to editors of medical journals, and footnotes in the stories about another footnote behavior, sneezing. I break with tradition and suggest that the photic sneeze is not a meaningless neurological quirk but joins stretching and yawning as a daybreak ritual, providing its daily nasal cleansing act. Although unexamined, photic sneezing may have a circadian pattern with the sun as *Zeitgeber* (time-giving stimulus) in

figure 7.2 *Sneezes produce powerful blasts of air that clear the airways, especially the nose. Sneezes can be triggered by allergens or mechanical stimuli in the nose* (upper), *irritants such as nasal snuff* (lower left), *or, in about a quarter of the adult population, bright light* (lower right). *Photic sneezes may be an involuntary daybreak response evolved to clear the nasal passages.*

susceptible individuals. Photic sneezing is an evolutionary experiment in which a novel stimulus (light) activates an already established motor pattern (sneezing). (The circadian act of yawning is most frequent upon waking but, unlike the sneeze, has another peak at bedtime. Neither is performed during sleep.) Sneeze stimuli appear in Fig. 7.2.

A past graduate student in my lab who was a photic sneezer observed that light never directly caused him to sneeze but would increase the probability of him doing so. If he felt a sneeze coming on, looking at a bright light would nudge him over the threshold. As noted previously, no one likes lingering on the threshold of either a sneeze or a yawn, a kind of body-wide holding pattern awaiting consummation.

Photic sneezes are not benign in our fast-paced modern environment. Vigorous sneezing triggered by driving through sunlit gaps in dense forest or emerging from a dark tunnel into bright sunlight can be a hazard. Sun-triggered sneezing puts aircraft pilots at even greater risk because of higher speeds and less margin for error. Even non-photic sneezes can be a risk; a high school friend of the author crashed the family car into a telephone pole during a sneezing fit.

Coughing and Sneezing Fetishes

Almost any stimulus can be linked to sexual arousal, including sneezing and coughing. The lesson is not the apparent weirdness of nontraditional sexual stimuli and those who fancy them but that we inhabit a body that is biased toward arousal and associated procreation. Your DNA does not care how the job gets done. The pursuit of fetishes is aided by the Internet, an uncensored source of enlightenment and entertainment for anonymous aficionados and researchers. Much of the material on these sites seems authentic. What prankster would catalogue all known sneezes or coughs in television shows, commercials, films, and the print media? My casual perusal of fetish sites identified more material about sneezing than coughing, and a different pattern of interest by their enthusiasts. Sneeze sites focus on the sneeze, while cough sites focus more on a variety of ancillary issues, such as coughs as a symptom, not on the cough itself. Sneezing and coughing fetishes will be considered together because both involve respiratory housekeeping tasks.

Coughing sites are slanted toward a presumably male audience and often feature cough-producing behavior such as smoking and disease, sometimes involving submissive female coughers. There is cross-over between smoking and coughing fetish sites. Videos on some cough sites were unexceptional except for their theme of naked, smoking women. Given the amount of exposed skin, discipline with lighted cigarettes is probably a priority for actors. The tobacco industry may be pleased that their product has appeal beyond nicotine. But it may be less pleased about an unsettling theme of some fantasies: beautiful women with prolific coughs, the nastier the better, and their decaying lungs. This disturbing imagery is not exclusive to contemporary websites. Consider the consumption chic of opera (*La Traviata*), and the fragile, swooning women of Gothic and Victorian fiction (*Wuthering Heights*) who succumb to the "damps" and night air. Are these expressions of high culture basically different from the fetishists' less artistic fare? Is solicitation of caregiving by women a latent theme? The bouncy babes of *Baywatch* are not the feminist ideal, but at least they look healthy and don't fall victim to tuberculosis or the deadly effect of an evening breeze.

For most of us, sneezing does not provide the pleasure of orgasm, but like coughing, it does have its admirers. I found no reports of sneeze-induced orgasms similar to the yawn-induced orgasms of some users of Prozac and related antidepressants. Web browsing found reports of women's "cute little sneezes," the attractive sneezes of a sexy male high school teacher, erotic stories based on the irresistibly arousing sneezes of a sexual partner, and the heart-rending breakup of a married couple bound only by sneeze lust. There are admirers of male and female sneezes, gay sneeze fetishists, and discussions of the ethics of causing nonconsensual sneezes in others. Allergies and cold symptoms take on a completely different meaning for these enthusiasts. A positive hedonic quality has not been established for sneezing, as it has been for yawning (Chapter 1), but both have similar properties, end in climax, and produce facial expressions in their early stages that, at least superficially, resemble those of an orgasm. In sneezing, yawning, and orgasm—but not in coughing—*interruptus* leaves the actor unfulfilled. I'm left wondering if some seventeenth-century aristocrats were getting more from their snuff than a nicotine high.

The science of sexy sneezes got a major boost in 2008 with the publication of "Sneezing Induced by Sexual Ideation or Orgasm: An Under-reported Phenomenon," by Mahmood Bhutta and Harold Maxwell in the *Journal of the Royal Society of Medicine*.[8] Their review included sporadic reports in the medical literature and anecdotes collected through a Google search of Internet chat rooms (search terms "sex, sneeze OR sneezing"). Among the many Internet participants, "17 people of both sexes reported sneezing immediately upon sexual ideation, and three people after orgasm." These two categories seem to be mutually exclusive—no one reported sneezing to both ideation and orgasm.

The journal's website posts eight responses to the Bhutta and Maxwell article from readers sharing their personal experiences with the phenomenon, most reporting sneezing to sexual ideation. Two of these individuals noted sneezing as a signal of their sexual interest in another person, one male relating an attraction to another male as a teenager, the other a male attracted to females. Within this population, sneezes are an honest (hard to fake) announcement of sexual interest, as are pupil dilation, penile or clitoral tumescence, and associated lubrications. The sneeze is involved in more than respiratory housekeeping. Borrowing from Mae West, you can ask a sneezing friend, "Are your allergies acting up, or are you just glad to see me?"

8

hiccupping

"Jennifer Mee, a 15-year-old who started hiccupping four weeks ago today and has yet to stop," was a story that led to her appearance on NBC's *Today*, and—according to Jennifer's family—fifty-seven calls from ABC's competing *Good Morning America*, plus contacts from *The Ellen DeGeneres Show* and many other print and broadcast outlets in the United States, Canada, and Great Britain (*St. Petersburg Times*, February 20, 2007). A Google search for "Jennifer Mee" brings up many pages of hits, and searching for "hiccup" yields additional hits for her. Jennifer is well established as the "hiccup girl." Whether spontaneous or psychogenic in origin, her experience documents the public fascination with hiccupping. Jennifer's bout of hiccups ended at five weeks, leaving her a bit sore, tired, and ready to move on, but compared to what others have gone through, her ordeal was short-lived and benign.

The demise of Pope Pius XII in 1958 is often attributed to intractable hiccups. Hiccups were probably only a symptom of his recurrent gastritis but may have resulted from or contributed to the strokes and pneumonia that were his ultimate cause of death. Chronic hiccups were certainly not lethal for Iowa farmer Charles Osborne, who holds the record for the longest bout of hiccups: more than sixty-seven years. His bout started in 1922 while he was slaughtering a hog and ended for unknown reasons in 1990. His affliction brought a strange celebrity, including a place in *Guinness World Records* as the man with the longest attack of hiccups, and an appearance on *The Tonight Show* with Johnny Carson. Despite his infirmity, Osborne led a normal life, running businesses, courting a second wife while hiccupping, and raising eight children before succumbing to complications from ulcers at ninety-eight years of age. Lest you think hiccupping is only a modern fixation, it has attracted the attention of Plato, Hippocrates, Galen, and other notables since antiquity.

There is a disparity between the great popular awareness and the meager scientific presence of hiccupping. If scientists and physicians have any interest in hiccupping, it usually concerns how to stop it, and sometimes the drugs or pathology that cause it. Cures for bouts of hiccups (a bout is defined as hiccupping on and off for up to forty-eight hours) is a lively area of folk medicine, and there are many suggested remedies, ranging from breath holding to eating sugar. Treatment of persistent hiccups (hiccupping for between forty-eight hours and one month) and intractable hiccups (hiccupping two months or more) has involved more extreme measures, from drugs to nerve crushing. (Authors disagree about the exact time criterion for persistent hiccups. Here and below, I use the standard provided by the cited authors.)

The most creative cure is offered by a physician, Dr. Francis M. Fesmire of the University of Tennessee College of Medicine, in a memorable paper in the *Annals of Emergency Medicine*: "Termination of Intractable Hiccups by Digital Rectal

Massage."[1] The approach of a rubber gloved Dr. Fesmire would disrupt any number of physiological patterns. His discovery earned him the gratitude of thankful patients and a 2006 Ig Nobel Prize. If Dr. Fesmire is unavailable, Dr. Majed Odeh and his colleagues at Bnai Zion Medical Center, Haifa, Israel, have confirmed his result and can provide their own digital therapy.[2] Fellow Israelis Roni and Aya Peleg suggest medicinal sexual intercourse, observing that ejaculation (but not other exertions) suddenly stopped a three-day bout of hiccups of one of their patients.[3] In their report, they did not mention masturbation, a more readily available therapeutic intervention that does not require a co-therapist.

In a 1932 medical review about hiccup, Charles Mayo of the famous family of physicians observed that the knowledge of a subject is inversely related to the number of treatments suggested and tried for it, adding that no disease has had more treatments and fewer results than persistent hiccups.[4] Little has changed in eight decades. The medical literature is full of case studies about novel causes of and treatments for hiccupping, but little about hiccupping itself. The present account addresses this neglect, treating hiccupping as a normal part of human nature, not merely a symptom, and considers its mechanism, development, evolution and pathology. Although the study of hiccupping dates from antiquity, it still offers research opportunities because of its position at the thinly populated frontiers of scientific and medical research.

~

"Hiccup" (or "hiccough") is onomatopoeia and acoustic shorthand for the behavior that produced it. The medical term is "singultus." The hiccup is a short, sudden inspiration that is terminated almost immediately by the closing of the glottis, producing the characteristic "hic" sound. The glottis also had

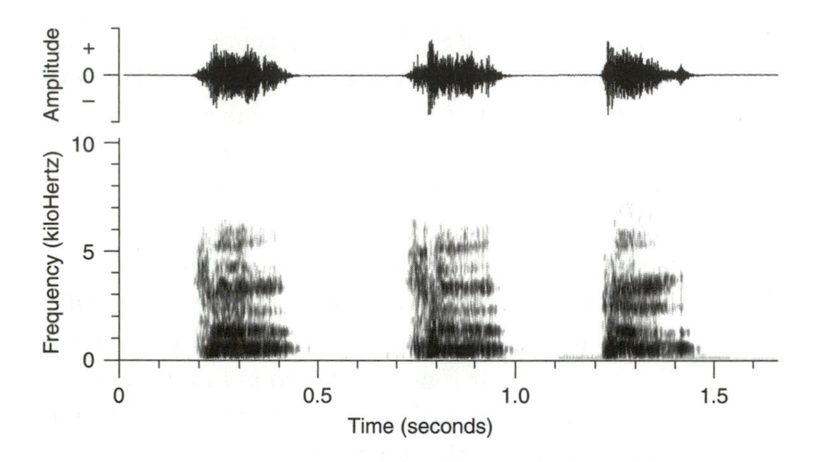

figure 8.1 *Waveforms* (upper) *and frequency spectra* (lower) *of three hiccups of an adult male. A hiccup is a sudden inspiration that is terminated almost immediately by the closing of the glottis that causes the characteristic "hic." The three hiccups are placed on the same axis for purpose of comparison; the intervals between hiccups do not reflect actual intervals between hiccups.*

a critical but different role in coughing and sneezing, which have been considered already.

The sonic and behavioral events of naturally produced hiccups are summarized in Fig. 8.1. The waveforms (upper) and frequency spectra (lower) show the initial inspiration terminated shortly after its onset by abrupt glottal closing. Hiccups are stereotyped in form and duration, and are shorter in duration than previously considered coughs and sneezes.

Hiccups are sufficiently common to be a well-known part of the universal human experience, but their unpredictable onset and low frequency make them difficult to study. As with sneezing, you can't simply go into the lab and get a study under way. Asking people to hiccup does not produce hiccups for study because they can't comply. But the failure to hiccup on command is itself informative, indicating that hiccupping is not under voluntary control. Our research team established this lack of control via reaction time. Subjects required an average time of 8.4 seconds to respond, demonstrating that they were essentially unable to hiccup on command. (The response latency probably would have been the maximum of ten seconds if not for a few game subjects who probably tried to comply by faking hiccups.) The latency of hiccupping was second only to crying (9.8 seconds), earning it a key on the far left side of our Behavioral Keyboard (in the appendix). In other words, we don't decide to hiccup—it just happens.

Although hiccups occur spontaneously, they can be influenced by higher-level psychological factors. This discovery occurred serendipitously while I tried to collect hiccups for study from the parade of students—one per hour—who pass through my wife's home piano studio. When a student started hiccupping, Helen signaled me to bring my audio recorder into the studio to record their sounds. The result was initially discouraging. In all nine cases, my appearance with recorder and microphone almost immediately brought an end to the hiccupping. In some cases, the mere threat to get the recorder stopped the hiccups. As noted elsewhere in this book, apparent failure is often success in disguise. One such result was the discovery of the just described "audio recorder cure" for hiccups. Actually, the recorder can probably be dispensed with—simply providing a witness to observe the hiccups would probably end a hiccup bout, at least with this group of young girls.

Of broader scientific interest is the principle of social inhibition of an unconsciously controlled, spontaneous act. Yawning is another unconscious act that is inhibited by social scrutiny (Chapter 1). When the ancient and the new, the unconscious and the conscious, compete for the brain's channel of expression, the more modern, conscious mechanism often dominates, suppressing its older, unconscious rival, whether hiccupping or yawning. However, preliminary evidence suggests that hiccupping is not suppressed by the more recently evolved act of speaking, as is laughter (Chapter 2); therefore, it does not punctuate the phrase structure of speech. Hiccups, indifferent to grammatical imperative, are distributed more randomly in the speech stream than is laughter.

Consider the previously mentioned case of Jennifer Mee, the "hiccup girl." In television interviews available online (at YouTube), her hiccups are atypical in punctuating her speech (most hiccups do not), evidence that they may be of psychogenic and not spontaneous origin. Her hiccups also sounded odd. Unfortunately, hiccupping is the least of Jennifer's current problems, as can be determined from Google. She has reappeared in the news, this time charged with murder, accused of luring a male victim to a robbery by accomplices that took a fatal turn. Her lawyer is planning a novel "hiccups defense," claiming them as a symptom (tic) of Tourette syndrome, which is offered as a mitigating circumstance. Aside from its questionable viability as a legal tactic, the hiccup defense requires the differentiation of normal hiccups from the similar vocal tic of Tourette. In contrast to normal hiccups, I suspect, a hiccup-like vocal tic of Tourette syndrome is produced by a different neuromuscular mechanism, sounds different, is more likely to punctuate speech, and is under more voluntary control. Hiccup science and Tourette syndrome may be on trial with Mee, promising a landmark case for all concerned.

~

Hiccups have attracted few dedicated researchers, but Terence Anthoney, MD, PhD, a professor emeritus of zoology at Southern Illinois University (Carbondale), may be the world authority.[5,6] Although he has a modest publication record, he has a vast store of unpublished research, anecdotal data, and case studies of a sort seldom found in the modern scientific and medical literature. Anthoney adopted an ethological approach to hiccupping, treating it as a typical behavior of our species instead of mere pathology. Some people are bird watchers; he is a hiccup watcher. He dealt with the rarity and unpredictability of hiccups by dogged persistence—more than forty years' worth, and counting.

In the most detailed of his two short, idiosyncratic articles, Anthoney surveyed *all* of the hiccups of twenty individuals during continuous periods ranging from three months up to eleven years.[6] He also collected questionnaire data and interviews about hiccups from fifty additional individuals. His long-term, longitudinal analyses of normal and abnormal hiccups are unique and suggest significant trends. In a telephone interview and emails, Anthoney shared his anecdotes, speculations, and insights about hiccups.[7]

Anthoney observed that the rate of hiccupping slows during a bout, with the inter-hiccup interval being shorter for hiccups earlier than later in a bout. A slow rate suggests that the hiccup ordeal is nearing an end. For many individuals, infants as well as adults, the number of hiccups in a bout is usually either small (fewer than four hiccups) or quite large (more than thirty hiccups). In two hundred consecutive bouts of hiccups by Anthoney's wife, his most studied and dedicated subject, she either had bouts of fewer than seven or more than sixty-three hiccups. This striking dichotomy of bout length suggests an interesting problem for a neuroscientist: with repetition, hiccups may entrain or program the nervous system such that they become increasingly more persistent and difficult to extinguish. Similar entrainment phenomena have been noted for epilepsy, migraines, and depression: the more often they have occurred, the easier it is to trigger a new episode. Evidently, the brain learns how to produce these conditions, and does so more easily with experience.

Hiccup bouts also tend to cluster, with one bout likely to follow another within twenty-four hours. Although hiccups can occur throughout the waking day, those subjects who showed circadian tendencies tended to hiccup most during the evening hours. Sleep stops hiccups in most people, but of course breathing continues during sleep, indicating that hiccupping and breathing are produced and controlled by different mechanisms, and that hiccupping is not simply an aberration of respiration. The different developmental trajectories for hiccupping and breathing considered below further support this conclusion.

Anthoney's patient research revealed an intriguing link between hiccupping and sex. Based mostly on over four decades of continuous observations of his wife through hundreds of menstrual cycles, Anthoney noted that her hiccupping was correlated with this cycle, with the most hiccupping occurring during the days immediately before ovulation.[6] Her hiccups essentially stopped during her two pregnancies but were present before and after the pregnancies. Thus, the hiccup is a potential signal of fertility and sexual receptivity of women. A search of the literature found no evidence that this window into female sexuality is recognized or exploited by males. Do you find hiccups arousing? If there's a hiccup fetish, it is neither common nor well known.

~

In his longitudinal study of a small group of subjects, Anthoney found that young individuals hiccup more often than older ones, and that young women hiccup more than young men, a sex difference established by adolescence and possibly much earlier. By menopause, most women have few bouts per year (perhaps zero to six), and men of similar age have even fewer (perhaps zero to two). Prompted by Anthoney's preliminary reports, my research team examined the development and sex differences of hiccups in a large-scale, cross-sectional study.

Our 465 subjects reported their age, sex, and days since their most recent hiccup. Hiccupping today, for example, gets a score of 1, the shortest possible latency. Latency is a measure of hiccup frequency, with a high rate corresponding to a short latency. Parents assisted in collecting data from their children, personally providing data for children seven years of age or younger. This simple approach provides a useful estimate of a highly variable, infrequent behavior, despite possible age and sex differences in memory and reporting. Data analyses are under way as this book goes to press, but here is a first look.

The frequency of hiccupping, as measured by the most recent hiccup, decreases with age, though there is a sharp decline in some subjects appearing around the age of twenty, later than the anticipated inflection point around the age of puberty. Overall, women hiccup more frequently than men, and we seek the development stage when this difference emerges. Again, the developmental inflection point may occur after puberty. Sex differences are not detected prenatally[8] and are minimal, if present, during infancy and early childhood. Sex differences emerge when there is a sharp developmental decline in hiccup frequency, suggesting that hiccupping is being inhibited, perhaps by androgens present in both sexes, with males having a higher titer.

These questionnaire data about age and sex are best considered in the context of two temporal bookends, the normal prenatal period at one extreme and age-related pathology at the other (Fig. 8.2).

Hiccupping is *far* more frequent before birth than at any other time of life. The presence of prenatal hiccupping is not a modern revelation. Women have long known that their babies hiccup in the womb, usually during the later months of

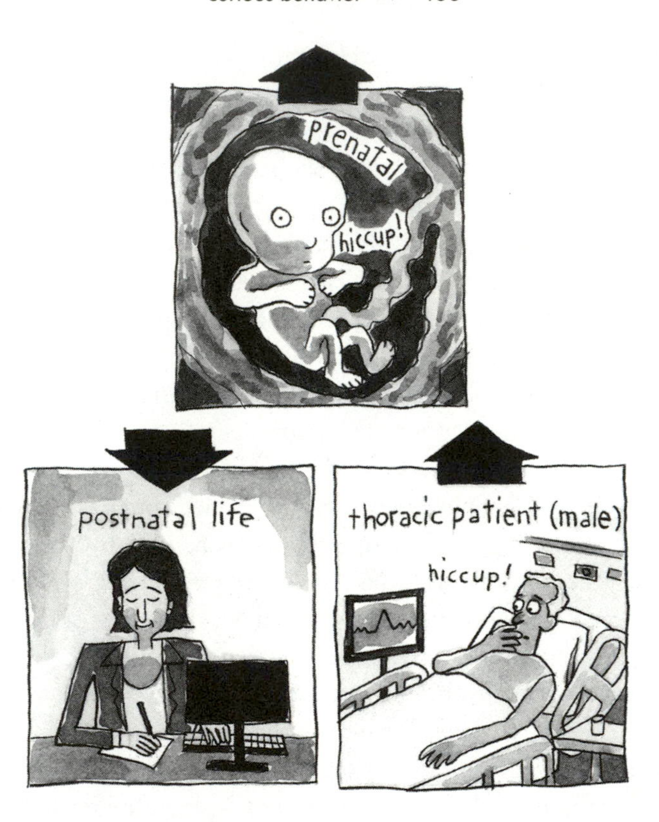

figure 8.2 *Hiccupping, one of the most common prenatal behaviors, peaks in ten-to-thirteen-week fetuses* (upper), *declines during postnatal life* (lower left), *but returns at high frequency in many male (but not female) thoracic surgery patients* (lower right).

gestation, when the hiccups are easier to detect. Women have told me of their delight with the "gentle," "cute" little rhythmic bounces in their belly, especially when they are able to identify them as hiccups. But fetal hiccups begin long before pregnant women can detect them.[9]

Using noninvasive ultrasonography, Johanna de Vries and colleagues in the Netherlands detected hiccupping in fetuses as early as eight weeks of gestation, during the first trimester of prenatal development, and only a week or two after the development of the diaphragm, the main muscle that produces the ingress (inhalation) of hiccupping.[8,10] (Full term is thirty-eight weeks.) Hiccupping quickly becomes one of the most common of all fetal movements, peaking between ten and thirteen weeks, after which it sharply declines until birth, a decline that, as described above, continues at a much lower rate during postnatal life (Fig. 8.2). These hiccups are not subtle—they involve high-amplitude bouncing and are so intense that pregnant women can sense them at late stages of gestation. After hiccups begin at around eight weeks, *the probability of a fetus hiccupping during a given day approaches 100 percent*, according to de Vries.[11] In contrast to the postnatal period, there are no sex differences in prenatal hiccupping. Breathing, another diaphragm-powered movement, also has an early onset during the eighth week, but it is initially performed at lower rates than hiccupping and increases in frequency during gestation while hiccupping is in decline.[8,9,12]

The very different developmental trajectories of hiccupping and breathing—breathing increases while hiccupping declines—suggest the involvement of different neurological motor pattern generators (oscillators). This is an instance where development provides insight into biological mechanism. The cessation of hiccupping but not breathing during sleep is further evidence of the independence of the two acts.

As a developmental neuroscientist, I immediately fixated on the early appearance, sharp increase, and subsequent decline of hiccupping frequency as birth approaches. This is a common trend in embryonic behavior and a central theme of this book's last chapter about prenatal development. As with other prenatal behaviors discussed there, hiccupping may have a function unique to the prenatal period, or none at all, with the act inhibited during the rest of the life span.

Let's shift now to a pathological phenomenon at the opposite end of the developmental continuum.

Intractable hiccups are experienced by only a few old people, but they are a clinically significant, age-related behavior that adds a reliable landmark to the often bewildering terrain of hiccup science. Individuals with intractable hiccups are overwhelmingly male and over forty years of age (Fig. 8.2). In a study of 220 cases of intractable hiccups from 1935 through 1963 at the Mayo Clinic, Jacques Souadjian and James Cain identified 181 males and 39 females, all who hiccupped more than two days, and most more than two months.[13] Of these patients, 93 percent of males were diagnosed with organic disease, while 92 percent of the females had psychogenic symptoms. Organic symptoms were most often medical (e.g., diaphragmatic hernia, arthrosclerosis) and postoperative. Psychogenic symptoms included addiction, grief, anorexia nervosa, and anxiety. Of males with non-surgery-related organic disease, most had other organic disease capable of causing hiccups. The high probability that pathological hiccupping has multiple causes indicates that clinicians should not stop their examination of patients after discovering a likely trigger. There may be others.

This clinical study reports an effect of a magnitude rare in behavioral science: men are *nine times* more likely than women to develop intractable hiccups.[13,14] This is a curious fact given that healthy, older adults of both sexes show relatively similar, low rates of normal hiccups. Although this sex effect is huge and reliable, its cause is unclear. Does the lowered testosterone of aging males disinhibit the latent pattern generator for hiccups? Is there something about male thoracic and abdominal anatomy that predisposes them to irritation that triggers hiccups? Or is hiccupping a symptom of organic disease such as arthrosclerosis that is more prevalent in males? The topic arose at a recent dinner party when a friend related a miserable experience with intractable hiccups during a hospital stay. Characteristically, he was male and in his late sixties, and his hiccups followed thoracic surgery.

Persistent hiccupping is associated with a variety of pathologies, including encephalitis, brain tumors, brain lesions, multiple sclerosis, meningitis, stroke, heart disease, kidney failure, diabetes, esophageal disease and irritation, gastritis, ulcers, diverticulitis, goiters, surgical operations, abnormal stimulation or irritation of the phrenic nerve, overeating, alcohol use, and various drugs, including midazolam (sedative), digoxin (heart medication), and steroids.[13,15,16] For an entertaining demonstration of hiccups caused by extreme esophageal irritation, see the YouTube videos of unfortunates who eat the fiery habanero pepper or the even hotter bhut jolokia, sometimes on a bet or dare. Swigging much milder Tabasco sauce may be a sufficient stimulus for many. This list of triggers is not complete, but it is sufficiently diverse to suggest that hiccupping has many causes and is, therefore, of limited use as a diagnostic symptom. Multiple triggers are consistent with the variety of medications used to treat chronic hiccups, including baclofen (a muscle relaxant), chlorpromazine (an antipsychotic), and metoclopramide (an anti-emetic).[17] Also, hiccupping may be initiated by positive or negative mechanisms; an example of the former is sensory stimulation by esophageal irritation, and an example of the latter is the release of a previously inhibited process caused by a brain lesion. Although there are a variety of triggers, all converge to produce the same motor pattern of hiccup.

Hiccup expert Terence Anthoney, as both physician and scientist, has studied and treated many individuals with intractable hiccups, including the record-holding chronic hiccupper Charlie Osborne. According to Anthoney, most individuals with chronic hiccups do not produce regular series of hiccups during their waking hours. Instead, based on data from fifty-five persistent hiccuppers or their physicians' reports, Anthoney found that they have cycles of several "on days" followed by several "off days," with a tendency for both on and off periods to lengthen with time. As noted previously, hiccupping usually stops during sleep.

In most patients with intractable hiccups, some of their hiccups become multiples (doublets are most common), occurring in rapid-fire bursts within a second or so. More than half of these multiple hiccuppers lapse into scary "tonic" hiccups that are sometimes mistakenly interpreted by physicians as spasms of the larynx.[18] These protracted super-hiccups involve all of the muscles that contract during a normal hiccup staying contracted for several seconds, sometimes up to thirty seconds or more. Imagine this experience! Patients are understandably frightened during the first episode of tonic hiccups because they are afraid that they will pass out or suffocate. From the patient's perspective, a tonic hiccup resembles the first stage of vomiting, with its powerful inward gasp, but this unpleasant state is sustained for many seconds. Fear diminishes when the patient discovers that the episode can be ended by a swallow of water, pounding himself on the back, or gagging himself with a finger. Vomiting and other reflexive acts can end hiccup bouts, with one of Anthoney's patients reporting that his hiccups would end with a rapid series of sneezes, a cure dating from Plato's time.

Using one robust act such as vomiting or sneezing to terminate another such as hiccupping suggests the involvement of a kind of neurological rebooting.

~

The most detailed description of the neurobehavioral events of the hiccup is provided by John Newsom-Davis of the Institute of Neurology, London.[19] Newsom-Davis tapped a reliable supply of hiccups: three patients seeking relief from intractable hiccups. Using electrophysiological recordings from muscles, he discovered fine details of the timing of muscle activation that are not detectable by visual inspection.

The hiccup begins with bursts of nearly synchronous activity in inspirational muscles (diaphragm and external intercostals) that lower interthoracic pressure, sucking air inward as when inhaling, but more abruptly. This inspirational surge is uncontested because muscles with an antagonistic, expiratory function (expiratory intercostals) are inhibited. The contraction of the main inspirational muscles may last half a second, but the duration of the inspiratory airflow is much shorter. The glottis slams shut about 35 milliseconds (.035 second) after the onset of a hiccup, blocking the rapid inhalation almost as soon as it starts, nullifying most of its ventilatory effect. Glottal closing occurs about 60 milliseconds (.060 second) before the contraction of the diaphragm reaches its peak. Without glottal closing, hiccupping would produce a large inflow of air.

The finding of inspiratory muscle excitation (diaphragm, external intercostals) while expiratory intercostal muscles are inhibited indicates the hiccup is neither a purely inspiratory event nor a massive, nonspecific motor surge involving the simultaneous contraction of all muscles. The hiccup is orchestrated by a neurological motor pattern generator that controls the timing and activation sequence of a variety of muscles.

Without the glottal closing that produces the "hic" in hiccup, hiccupping would produce a large inward flow of air, as it did in one of Newsom-Davis' patients with a tracheostomy.[19] The modified airway of this patient bypassed the glottis, producing severe and potentially dangerous hyperventilation. Fred Plum, Newsom-Davis' Cornell-based mentor in respiratory neurology, noted a "patient with brainstem infarction and a tracheostomy in whom sustained hiccup provided enough air to ventilate him for many hours."[20] It's unclear whether Plum and Newsom-Davis are talking about the same patient. However, both observations establish the respiratory potential of hiccupping and suggest a respiratory heritage for the act.

Having described the motor pattern of the hiccup, Newsom-Davis examined factors that influence it. He discovered that hiccups occur much more often during inspiration than expiration and that breathing an enhanced level of carbon dioxide (5 percent carbon dioxide, 95 percent oxygen) decreased the frequency but not the amplitude of hiccups. The inhibitory effect of carbon dioxide on hiccup frequency is consistent with the folk remedy of holding one's breath or rebreathing air in a paper bag to stop hiccups. The effect of carbon dioxide on the frequency but not the amplitude of hiccups is further evidence that the two variables are independently controlled. In Chapter 1, carbon dioxide inhalation was shown to influence the rate of breathing but not yawning, establishing the independence of the two acts and rejecting the hypothesis that yawning is a response to a heightened level of carbon dioxide.

I like the late Newsom-Davis' catch-as-catch-can, small-sample approach to a difficult research problem and his willingness to use himself as subject.[21] He once gave himself a pneumothorax while attempting to record electrical activity from his own intercostal muscles or diaphragm, and described his deflated lung as "a cold fish flapping around in his chest." He recovered spontaneously and became a leading clinician, scientist, and scholar.

~

The hiccup serves no obvious function in human postnatal life. Explanations of hiccup function typically consider it a developmental[22] or evolutionary vestige,[23] referring to its presumed functions in prenatal life or in phylogenetic antiquity. For example, is hiccupping a remnant of gill breathing of our aquatic ancestors, or of a motor pattern generator that once contributed to the development of other rhythmic acts such as sucking, gasping, breathing, or embryonic skeletal and neuromuscular development? Or does it have no function at any stage of life?[24] Although these developmental and evolutionary issues are interesting, they are also necessarily complex and speculative. A more pragmatic, if less adventurous, initial approach to hiccupping is to study the facts of its development, production, and control, both in health and in disease.

This chapter demonstrates that aspects of hiccup science are open to anyone and easy to get under way. Consider the previously discussed developmental study that simply asked males and females of different ages when they last hiccupped, the test of voluntary control of hiccupping via reaction time, the discovery of social inhibition by trying to record hiccups, and Terence Anthoney's long-term studies of his family and friends. Not whiz-bang science, but a start and a solid foundation.

The most fundamental research is yet undone, including detailed neurobehavioral descriptions of hiccups before and after birth in humans and other animals, in health and disease. Such accounts define the behavior to be explained and have a long scientific shelf life: Newsom-Davis' neurophysiological study of three patients remains the state of the art more than four decades later. The simple, species-wide, stereotyped, rhythmic character of hiccups makes them easy to describe and ideal for rigorous neurological analysis. Unfortunately, these virtues of hiccups are diminished by the low frequency and unpredictability of the act. Trapping the wild hiccup is not easy. The problem of waiting for hiccups that never come can be partially solved by observing individuals with high rates of hiccupping, such as fetuses and adults with intractable hiccups. The full power of neurological analysis awaits the development of an animal model that reliably produces hiccups in the laboratory. The ability to observe neurological circuits in action would revolutionize hiccup science and solve age-old problems, perhaps including the evolutionary origin of this elusive act.

~

I end this chapter as I opened it, with evidence of the great public interest in hiccupping and its cures, this time from twenty-three hundred years ago. The source is Plato's *Symposium* (literally a "drinking together"), a fictional account of philosophical dialogue at a dinner party at which the all male guests expound on the nature of love. Aristophanes' turn to speak was delayed by a bout of hiccups. (Aristophanes was spared the more scatological afflictions of his own theatrical characters.) The bombastic physician Eryximachus (literally "battler of belches"), the next speaker in line, agrees to speak instead, and suggests a cure to Aristophanes: "Hold your breath, and if after you have done so for some time the hiccup is no better, then gargle with a little water, and finally tickle your nose with something and sneeze; and if you sneeze once or twice, even the most violent hiccup is sure to go." Aristophanes later reports that "sneezing finally worked." It is humbling to recognize that these cures are still recommended and that despite contemporary discoveries, hiccupping remains a mystery. Perhaps the next twenty-three hundred years will bring more progress.

9

vomiting and nausea

My first serious thinking about vomiting (emesis) and nausea came at age twelve during a cross-country family automobile trip. Six family members piled into our big green Buick sedan, and we set off on our long journey, planning to save time and money by making ham sandwiches from ingredients stored in an ice chest instead of stopping at a restaurant. The flaw in this plan became apparent several hours after our lunch on the road. Cousin Karen started to feel queasy and soon started vomiting, only sometimes managing to discharge out of the car's window, a significant oversight. The sight, sound, and especially the smell of a violently retching Karen were most unsettling. Soon everyone in the car was starting to feel nauseated, and several more barfed, amplifying the effect. Everyone, including me, ended up in a motel room, joining in an inglorious, evening-long session of communal puking. It's amazing how your universe shrinks when you are kneeling before a toilet bowl waiting to vomit. Decades have not dimmed the lessons I learned that evening.

My first, rather obvious conclusion was that vomiting (the reflexive, forceful ejection of stomach contents through the mouth) and associated nausea (the sensation that one is about to vomit) must be very biologically important to hijack our behavior and consciousness so convincingly. More novel was the revelation that nausea/vomiting can be social acts, explaining why one vomiting person triggers vomiting or at least queasiness in witnesses. (Although I, too, probably was poisoned, I suspect that I was a borderline case and might not have succumbed had I not been influenced by the actions of others.) The contagious response was powerful and automatic, certainly not involving my desire to imitate a model. If you doubt the contagiousness of vomiting, talk with school janitors who hasten to clean up one child's vomitus before it triggers a storm of sympathetic responses from highly suggestible classmates.

Contagious vomiting and nausea are more than neurobehavioral flukes and housekeeping nightmares. The first vomiting individual serves as the communal taster whose behavior warns other group members, who respond contagiously, preemptively clearing their digestive systems before they experience symptoms. Messy false alarms are a modest price for a potentially lifesaving maneuver—it's better to be safe than sorry.

Vomit joins other body substances such as feces, urine, blood (especially menstrual blood), and decay as the type of substance and odor for which we might have evolved innate rejection.[1] All are possible vectors for disease and have been reviled throughout history and across cultures, but, surprisingly, there is no evidence of innate aversions to nonirritating substances and odors.[2] Young children do not reject feces—rejection of decay odors does not appear until between three and seven years of age and probably is associated with toilet training, which can start around two years of age. There is no evidence of instinctive aversions in nonhuman animals. Although the aversion responsible for contagious vomiting in humans is not innate, we have evolved the learning process through which such aversion is acquired.

~

The topic of vomiting and nausea is rich at many levels, affecting even our vocabulary, an observation that should be obvious to those finding this discussion nauseating or disgusting (the latter meaning, literally, "bad-tasting"). "Nauseating" and "disgusting" have become such vivid adjectives that they are used to describe unpleasantness in everything from bad mackerel to bad politicians. Vomiting has a firm place in popular culture, as reflected in its rich and growing list of synonyms, including "throw up," "upchuck," "barf," "puke," "heave," "spew," "hurl," "retch," and "toss your cookies." These terms somehow escaped the founders of the Chuck-A-Burger chain of drive-in restaurants, which I recall from my student days in St. Louis.

Vomiting has a complex muscular choreography orchestrated by neurons in the brain stem.[3] Vomiting begins with one or more violent contractions of the diaphragm and muscles of the abdomen and chest, which put pressure on the stomach. This intense spasm, the retch, produces a powerful inward gasp that resembles a violent hiccup, with which it's sometimes confused and to which it may be related. The deep inspiration against a closed glottis, along with increased abdominal pressure, produces a pressure difference between the abdominal and thoracic cavities that propels gastric contents upward. The expulsive phase of vomiting is triggered by the sudden release of pressure in the abdomen by the relaxation of the upper esophageal sphincter, which causes the contents of the stomach to be ejected through the esophagus and mouth. The stomach does not contract and the esophagus does not perform reverse peristalsis during vomiting.

The termination of vomiting brings blessed relief from cramping abdominal muscles and nausea and may even produce a squirt of endorphin. Dry heaves (unproductive emesis) bring only pain and continuing disability. If aspects of vomiting behavior seem vaguely familiar, it's because we have previously encountered the biological tactic of damming, pressurizing, and then explosively releasing body contents in coughing (Chapter 6) and sneezing (Chapter 7). Useful neurobehavioral mechanisms are recycled during evolutionary history.

Vomiting is not regurgitation, a more passive, less unpleasant act lacking abdominal contractions in which undigested stomach contents flow back into the esophagus and sometimes out of the mouth, as when babies spit up after feeding, sometimes accompanied with a burp. The food comes up before it has gone down, more a spilling than an ejection. Babies who are overfed or not held in an upright position are more likely to pour ingested substances from their mouth, a situation that resolves naturally with the maturation of the sphincter that keeps stomach contents from reentering the esophagus. Regurgitation lacks the sensation of nausea and can be an unwelcome accident that comes as a surprise. Specialists distinguish the products of vomiting and regurgitation by sniffing—a baby's spit-up smells like sour milk, not vomit. Sniffing, probing, and visualizing the body's ejecta (from both ends) were state-of-the-art medical tests of an earlier era.

~

"Over the lips and through the gums, look out stomach, here she comes" goes a Rabelaisian toast. Eating and drinking are commitments. We are what we eat, and we must regulate eating and drinking to grow and stoke our metabolic fires. An unromantic view of the human being reduces us to a tube with sharp teeth at one end and a sphincter at the other. Entry to this tube is guarded by our dual chemical analysis systems, the senses of smell (olfaction) and taste (gustation), as well as tactile, irritant, and general senses in the mouth. As any desperately hungry person knows, when retrieving the contexts of a mysterious foil-wrapped parcel from the back of the refrigerator, it's best to sniff before tasting. Is it guacamole or last week's spaghetti?

Sniffing requires less of a biological commitment than tasting, although we do have a more intimate relation with odors than you may imagine. The sensory surface in the upper recesses of the nasal cavity (the nasal mucosa) acts as a tiny chemosensitive "tongue" that "tastes" molecules of substances wafted into solution on its mucus surface—a touching insight when you detect the scent of a loved one, less so when you smell shit.

The stakes are so high in matters of eating and drinking that we are born preprogrammed to nibble our way successfully through the buffet of life, instinctively choosing proper foods, stopping when full, and varying our diet. Our crude dietary predispositions acquired through natural selection are fine-tuned through experience. Generally, substances that taste good are nutritious, and those that are unpleasant are not. Sweets, for example, are irresistible because they are a concentrated source of energy-rich carbohydrates. Newborns seek sweet substances, which are usually safe and nutritious, and turn away from potentially dangerous bitter substances.[2,4] A taste for coffee, a mildly bitter substance, or jalapeño peppers, a painfully burning substance, is acquired later. Newborns seem to lack equivalent, innate preferences and aversions for non-irritating aromas.

There are brakes on our consummation of even desirable foods. We feel sated and stop eating long before we burst, unlike the gluttonous diner in *Monty Python's The Meaning of*

Life who succumbed to that one last ever-so-thin mint. The fruit fly (*Drosophila melanogaster*) really can eat to the point of near bursting if the nerve between its gut and brain is cut, so that fullness information can't be used to terminate eating.[5] We humans, unlike the fruit fly, have redundant systems controlling food intake, and except in extreme cases, our sensations of satiety may be food-specific. A chocolate binge, for instance, may not eliminate our interest in fruit, bread, or steak. Our biology is biased against starvation, a more immediate health problem than the long-term effects of obesity. It's understandable that overeating has become the most common eating problem in our current time of plenty. However, extreme, short-term pathologies of eating have also emerged. Bulimia nervosa is the loss of control of food intake, mostly by women, characterized by bouts of excessive hunger and eating followed by forced vomiting or purging with laxatives, commonly accompanied by feelings of depression and guilt.[6] Ironically, bulimia is sometimes coupled with the dangerous condition of anorexia nervosa, the exaggerated concern with being overweight that leads to excessive dieting, often compulsive exercising, and sometimes fatal starvation. In bulimia, the potentially lifesaving maneuver of vomiting becomes pathology.

All is not lost if we eat something toxic or disgusting. We can spit out what doesn't suit us, and vomit what has already been swallowed, a direction opposite the vivid gustatory imagery prompted by Italian dishes called *saltimbocca*—so tasty that they literally "jump into your mouth." Vomiting can be lifesaving for species that can do it, permitting a second chance to make a better dietary choice. Rats and other creatures unable to vomit can learn to avoid food that sickens but does not kill them.[2] They may also resort to pica, the eating of non-nutritive substances such as clay that may buffer the effects of toxins. Like us, they develop powerful, long-term aversions to foods that don't kill them. The blood thinner warfarin is an effective rat poison because the lethal internal hemorrhages it produces build so slowly that rats don't get sick and acquire an aversion to tainted bait.

The details of food avoidance are significant beyond the community of eating experts and rat exterminators—they force a rethinking of a major scientific issue, how we form a conditioned response.[7] Psychological dogma once dictated that the formation of a Pavlovian (classical) conditioned response required a short interval of around one-third to two-thirds of a second between the occurrence of the conditioned stimulus (in this case, the taste of food) and the unconditioned response (sickness). However, as anyone who has ever eaten tainted food knows, aversions can be acquired to foods sampled hours earlier, a latency necessary for bad food to make us sick. Food aversions can be established in only one trial and can be long-lasting. I avoided Burger King restaurants for five years after becoming ill after eating what I thought was a tainted Whopper.

The food avoidance response is selective, with a preparedness to form some associations more readily than others, especially those tastes associated with toxic agents or disease.[8] We can, for example, learn to avoid food-related tastes and smells that precede nausea, but not sounds, lights, flowers, or machines that have nothing to do with food. Conditioned cues for nausea can be remote from the initial experience of sickness, as when a cancer patient becomes ill in a clinic parking lot in anticipation of chemotherapeutic injections that have previously produced nausea.

The substances that we find disgusting and perhaps nauseating were examined by Paul Rozin and colleagues at the University of Pennsylvania.[2] For adults, feces seem to be the most universally reviled substance, with the odor of decay being the most disgusting sensory attribute. Both make a lot of biological sense because they are vectors of disease. But when it comes to food, we are not always coldly rational consumers. For example, dipping a perfectly harmless, germ-free, sterilized dead cockroach in a beverage renders it undrinkable for the North American college students who were Rozin's subjects.[9] And perfectly good fudge was deemed undesirable when in the shape of a dog turd. (The attractiveness of turd-shaped fudge on a lawn or, conversely, square, fudge-shaped turds on a dining room table went unexplored.) Such disgust is produced by superficial, "magical" properties of substances, such as contact (sterile cockroach) or shape (turd-shaped fudge). Harmless body products become disgusting after leaving the body, as when spitting into a glass of water that you are about to drink.[2]

Disgust can even generalize to sounds. In an online study of more than a million responders, Professor Trevor Cox of the University of Salford (United Kingdom) found vomiting to be "the worst sound in the world." Visit the Bad Vibes section of his website (www.sound101.org) to participate in his study and to sample his stimuli.

During a recent dinner party for our friends Doug and Kay, my wife and I unintentionally explored the palatability of foods that smell like vomit. When seated at the table, Kay had an unsettled expression and requested that one of our cheeses be removed because its odor was making her ill. Although I found it tasty, that aged Spanish blue was one funky cheese. Its odor was so strong that I kept it carefully wrapped in the refrigerator and avoided touching it with my fingers because its odor would linger after several washings. In retrospect, such ripe, soft cheeses smell strongly of vomit, an observation confirmed by our connoisseur of disgust, Paul Rozin. Give a ripe cheese a sniff and draw your own conclusion. Cheese is one of the select, culture-specific, smelly foods based on fermentation and decay, including meat for Inuits and fish for Southeast Asians (fish sauce).[2] Not all people are attracted to the stinky, fermented, coagulated secretions of modified sweat glands that we know as cheese, and many find it disgusting.

This chapter describes the sensation of nausea and act of vomiting wherever they occur, whether or not they are food-related. This is significant because nausea and vomiting can be produced by stimuli not directly related to food or infectious disease, as with motion (sea, air, space) sickness.

Motion sickness has been known since humans took to the sea in ships, with the word "nausea" derived from the Greek word *nautia,* "seasickness." Ironically, the word "sea" is in nausea. Space sickness is a modern manifestation with its own unlikely legend, Edwin "Jake" Garn, a former Republican senator from Utah who chaired the Senate subcommittee that controlled the NASA budget. After bullying a ride on the space shuttle *Discovery,* Astronaut Garn set a standard for space sickness, memorialized jokingly in the Garn Scale, in which one Garn is the "mark of being totally sick and incompetent. . . . Most will get maybe to a tenth [of a] Garn, if that high," during their early days in orbit.[10] The probable cause of space sickness, like car sickness, is a conflict between information from your body's motion detectors (semicircular canals and otolith organs of the vestibular system) that you are moving with information from your eyes that you are standing still; whether in a space shuttle or a car, the walls of your container are not moving.[11] Disorientation is particularly unsettling in space because, lacking the sensory anchor of gravity, you don't know which way is up or down. Astronauts feel nauseated until they recalibrate their sensory systems. Susceptibility to space sickness can't be accurately predicted from Earth-based tests because of the ever-present pull of gravity, but the drugs promethazine and scopolamine may reduce its symptoms.

figure 9.1 *The driver of a boat or other vehicle is less likely to become nauseated than a passenger. A boating example is fitting, given the derivation of "nausea" from the Greek word* nautia, *"seasickness."*

Passive movement is the source of most motion sickness. Seasickness, for example, is caused by the passive motion of waves that are beyond your control. The similar cyclic up-and-down stimulation of the active motion of walking does not cause walking sickness because the brain cancels out the consequences of self-produced movements. Motion sickness is lessened for an individual who has some active control over the movement (Fig. 9.1). The driver of a car, for example, is less likely than a passenger to become carsick. As car-driving, seafaring, space-traveling apes, we are not well served by our vestibular system, a movement detection apparatus evolved to serve a self-propelled walking or running biped in a three-dimensional environment subject to Earth's gravity.

It's not clear why a stimulus such as motion that has nothing to do with food causes nausea and vomiting. An untested explanation by Michel Treisman of Oxford University is that ingested toxins mimic symptoms resembling those of movement sickness, and we protect ourselves by vomiting.[12]

~

NEWS FLASH: Elementary school students from East Templeton, Massachusetts, visiting a local high school to practice for their annual spring choral concert, suffered an outbreak of fainting, nausea, and respiratory distress (1981).[13]

Illnesses at Anchorage, Alaska, junior high school attributed to sewer gas (1987).[14]

Group of seventh- and eighth-grade students from Ontario were overcome in a Montreal train station on their way home from a four-day cultural exchange visit to Quebec City (1981).[15]

Alabama high school marching band experienced fainting from exhaust fumes in auditorium (1973).[16]

Incidents like these will be reported in the news many times this year. One occurred while I was writing this chapter: "More than 20 people, mostly band members, were hospitalized after suddenly falling ill during a high school football game in southwest Houston" (FoxNews.com, November 4, 2011). You can immediately recognize them by the demographics—most involve groups of young people, especially girls. The symptoms will be vague, most often being nausea, faintness, dizziness, shortness of breath, or "feeling strange," and the cause will be undetermined, frequently attributed to gas or funny smells. The demographics and scenario (unspecified cause, children but not adults affected, more girls than boys affected, group situation and membership) are

figure 9.2 *The sight and smell of a vomiting individual can trigger contagious, psychogenic responses in witnesses, especially girls of middle-school age, and others during stress-producing social events such as school band and chorus trips.*

tip-offs that an episode is attributable to psychogenic illness. Susceptibility to such illness is increased by anxiety, stress, heat, crowding, and sleep deprivation, conditions especially present on school trips and outings. The appearance of emergency responders wearing gas masks and news media covering the crisis validates the rumors and further escalates the anxiety of the affected population.

During a 1979 episode of mass hysteria in a Boston suburban elementary school, thirty-four of 224 students attending a play honoring graduating sixth-graders were hospitalized for severe dizziness, weakness, hyperventilation, headaches, nausea, and abdominal pain.[17] Nausea was reported by 44 percent of the hospitalized students (Fig. 9.2). An estimated forty to fifty additional students reported less severe symptoms but were not hospitalized. The episode was triggered by a sixth-grade boy, a student leader, who experienced dizziness, fell from the stage, cut his chin, and bled profusely. Rumors about the outbreak spread rapidly through students and the outside community. Two priests arrived to attend to friends and family of the twelve children presumed to have died of food poisoning. Fortunately, the "deceased" students didn't require last rites.

For a geopolitical twist, consider the 1983 report of a "mysterious gas poisoning" of nine hundred people, mostly schoolgirls, on Jordan's West Bank.[18] Both Arabs and Israelis believed that a poisonous substance was involved, an interpretation not discouraged by the uncooperative staff of the Jordanian hospital. In the United States, a few weeks after the start of the Persian Gulf War in 1991, there was an outbreak at a junior high school in Central Falls, Rhode Island.[19] A precipitating factor may have been anxiety about the Gulf War and the associated threat of gas, chemical, and bomb attacks on the U.S. mainland. At a time of high anxiety, it did not matter that Central Falls, Rhode Island, was an unlikely target of the late Saddam Hussein's weapons of mass destruction. In May 2009, about two hundred Afghan girls fell ill after smelling perfume, flowers, rotting garbage, or cigarette smoke.[20] Some collapsed, while others reported headaches, dizziness, vomiting, and stinging in the eyes. All recovered quickly and had no medical evidence of toxicity. Some Afghans, including officials, feared that the Taliban had discovered a new means for keeping girls out of school—poison.

In many cases of mass psychogenic illness, the first person of a group to be affected, the *index case*, may experience actual physiological illness. Others may experience a sympathetic, psychogenic response, especially in the presence of facilitating factors such as fatigue, strange smells, disorientation, or simply not feeling quite right. Our brain is scrambling to find a cause of our illness, however tenuous the link. The index individual is our unappointed communal taster, a sentry who provides early warning about the safety of our food, drink and environment. We are beneficiaries, not victims, of this adaptive quirk in our sociobiological programming.

It makes good evolutionary sense that smells ("gas") are a regular factor of stories about psychogenic illness. Olfaction is the body's chemical vanguard, our hair-trigger detector of toxic threat, real and imagined, that initiates a massive, if crudely targeted and often ineffective, group response. Bystanders experiencing ambiguous feelings may find an explanation in the faintness or nausea of the index case. Psychogenic illness feels no less distressing to the person experiencing it and can lead to real physiological consequences. This position is hardly controversial—many people experience a sympathetic response to viewing a bloody accident victim, a seriously ill or disfigured person, or, as in the case of my childhood auto trip, nausea and vomiting. Businesses as well as school and public health officials must contend with the challenging managerial and financial consequences of this human herd behavior.

Consider the Coke recall in Belgium in June 1999 that cost the company more than $100 million in expenses and damaged the brand image.[21] The outbreak started on June 8, 1999, when ten students at a secondary school were admitted to a hospital after drinking Coca-Cola. Symptoms included abdominal discomfort, headache, nausea, vomiting, malaise, respiratory distress, trembling, and dizziness. A rapid inquiry by school staff found that all victims had drunk Coke, a possible source of the illness. The Cokes were reported by affected students to have an abnormal odor and/or flavor. The staff then questioned students in other classes, asking who had drunk Coke and was not feeling well. This investigation was followed by a second wave of illness that evening and following morning. A total of thirty-seven children between the ages of ten and seventeen (with a mean age of thirteen) were admitted to the hospital, with a susceptibility rate for girls of 16 percent (28 of 179) and for boys of 9 percent (9 of 101). The Coke story was covered by television news bulletins, and

Coca-Cola, in a press release, admitted problems with product quality and stated that there was no health risk.

Over the next few days, students from four other schools in different parts of Belgium became ill after drinking Cokes and were taken to hospitals, usually by ambulance, and often accompanied by television news crews. The seventy-five children, who ranged in age from thirteen to nineteen (with a mean age of fourteen), had a susceptibility rate of 4 percent for girls (72 of 1,666) and 1 percent for boys (3 of 394). (The epidemic extended to France but public health data are not available.) "Poisoning by Coca-Cola drinks" was now a national issue, and on June 14 the Belgian Health Ministry forced the total recall and destruction of all Coca-Cola soft drink products despite the absence of measurable contaminants or a toxic reaction in the sick students. On June 23, satisfied that the crisis had passed and concerns about toxicity had been addressed, the health authorities permitted the sale of Coke products to resume.

Although there may have been minor odor and taste issues with Coke products and containers from one of several bottling plants, the main problem was a human herd reaction coupled with a slow, inept corporate response. There were also premeditating factors. Belgium was still reeling from the "dioxin crisis," in which a leak to the media on May 25, 1999, revealed that the food supply was contaminated by a dangerous carcinogen. The ensuing food scare resulted in a recall of eggs and chicken, and later most meat and dairy products. The ministers of health and agriculture were forced to resign. Belgian citizens were primed for a food-related psychogenic event by a barrage of worrisome media reports about the dangers of even trace amounts of toxins coupled with a loss of faith in government oversight. In these circumstances, the whiff of an odd-smelling Coke, real or imagined, was enough to push a vulnerable population of young students and a few adults into psychogenic illness. The newly appointed minister of health was also highly motivated to act in the public interest given the recent sack of his predecessor. It did not help that Coca-Cola was an iconic foreign corporation and that its leading product has a famously secret formula.

Other companies face similar, difficult choices, the stuff of business school case studies. It's wise for a business to have an action plan ready when it faces its own crisis. There is no painless solution; the best tactic is the least bad option. Foot-dragging and the decision not to recall products send an ill-advised and ultimately costly message that the company values the bottom line more than consumer health and goodwill. In most cases, the only viable choice is a recall or closing, with the hope that public attention and the associated psychogenic effects will soon abate, and follow up with a vigorous public relations campaign. Many hotels, businesses, and schools occupying so-called sick buildings—presumed to have toxic gases, faulty air-conditioning, or mold—may be burdened with costly, difficult-to-solve perceptual problems. Imagine the challenge of middle school and high school administrators who ride herd on the perfect incubators of psychogenic illness, large groups of children of the most susceptible ages. Without becoming insensitive to legitimate complaints, managers must understand the difficult, politically delicate, and costly dilemma of being forced to seek physical and medical solutions for what is often a psychological and social problem. There is also the possibility that a presumed psychogenic outbreak may really have a physical basis.[22] These complications highlight that even the apparently physiological phenomena of vomiting and nausea occur in social contexts.

~

Vomiting makes us feel terrible, and it's inconvenient, expensive, messy, and sometimes dangerous. A lot of people want to understand it and make it stop, whether medical professionals or those unfortunates kneeling before the porcelain urn. Research on vomiting and nausea is presented in many medical and scientific settings, including the huge annual meetings of the Society for Neuroscience that I attend. To provide a more personal touch to this scientific extravaganza of thousands of presentations and more than thirty thousand attendees, the society organizes small special-interest socials and dinners around narrow themes. Past meetings included Club Capsaicin, whose members visited area Thai, Mexican, Indian, or Chinese restaurants to sample the acquired taste of cuisine made hot by the capsaicin in peppers. Club Emesis (vomiting) was another special-interest group, but I have no record of its outings. It seems fitting, though unlikely, that these connoisseurs would celebrate their specialty by raising their glasses to drink a toast of ayahuasca, a ritual emetic used by some Central and South American peoples who seek purification and bonding through communal vomiting.[23] The highlight of the evening could be an emesis film festival, featuring excerpts from such disgusting fare as *Monty Python's The Meaning of Life*, *The Exorcist*, and a personal favorite, the "Legend of Lard Ass" scene from *Stand by Me* that concludes with a crescendo of contagious vomiting.

10

tickling

You cannot tickle yourself. From this basic observation comes insight into the neurological program for social play, the neurological computation of self and other, a possible defect in autism, and how to program personhood into robots and improve their performance. Insights that now seem so clear had an uncertain genesis.

The impetus for my program in tickle research came via astronomy. I was invited to speak about laughter at the Goddard Space Flight Center, an exciting opportunity for an amateur astronomer and telescope builder such as myself. Before my talk, a young female astronomer volunteered that she certainly hoped that I wasn't researching tickling—she hated being tickled. Her revulsion was persuasive. A behavior so aversive to some yet so attractive to others, especially young children, must be important. I immediately started to examine who tickled whom and why. No grant funding was necessary. Unlike the big science of the Goddard astronomers, the frontiers of tickle research can be pursued by anyone, regardless of resources. This is a good thing. Tickling lacks scientific gravitas, and its researchers are unlikely to earn either funding or a Nobel, although an Ig Nobel Prize is a definite possibility.[1]

Tickling involves at least two people, the *tickler* and the *ticklee*; therefore the tickle is inherently social. A good starting point for tickle research is to identify the players in the tickle game. This is easy because you already know these people very well. When have you ever tickled or been tickled by a stranger? Nonconsensual tickling by a stranger is an unwelcome intimacy, almost never happens, and may even be illegal.

Overwhelmingly, you tickle and are tickled by friends, family, and lovers, according to questionnaire responses that I obtained from 421 males and females between eight and eighty-six years of age.[2] Tickling is a form of nonverbal communication between significant others, not a reflex that occurs independent of social context. As an innate response to a tactile stimulus, tickling certainly has some reflex-like properties, but its complexity, duration, and sociality are not typical of classical reflexes such as the patellar tendon (knee-jerk) reflex.[3]

The rationale given most often by my respondents for tickling someone is "to show affection," followed by "to get attention."[2] Being held down and tickled by an older brother till you pee in your pants (an anecdote volunteered by a few respondents), though not rare, is an atypical scenario for most everyday tickle experiences. It does, however, suggest the vigor of the tickle response, a body-wide physiological eruption that can leave the recipient breathless, with heart racing, and, in a few cases, with damp underwear. Paradoxically, ticklees often punch and kick wildly in their desperate struggle to escape and stop the stimulation, only to return for a second dose of tickling.

figure 10.1 *Tickling is neurologically programmed social play that binds us together in the give-and-take of childhood rough-and-tumble and adult sex play. Tickling is not fun for the tickler if the ticklee does not respond, and it is not fun for the ticklee if he or she can't reciprocate. Tickle is necessarily social because you can't tickle yourself. Feigned tickling is probably the most ancient joke.*

Reciprocity is an essential element of tickle battles, the primal basis of physical play (Fig. 10.1). In tickle battles, tickler and ticklee are bound together in the laugh-filled give-and-take that may be the most benign form of human conflict. Consider the social choreography of tickling. The ticklee may huddle, push away the offending hand of the tickler and escape, only to return, renew the interaction, and counterattack. "I like to tickle someone only to get them back when they are tickling me," noted a twenty-one-year-old female respondent. "I liked tickling my brother to get him to chase me and play tag when we were young," observed another.

Tickle battles are instinctive social play that binds tickler and ticklee together in a primal, neurologically scripted encounter.[2] The tickle game is part of our mammalian heritage; reptiles and social insects don't participate in such tussles. Watch young squirrels at play, racing through the branches, ending in tumbling frolics in the leaves that resemble human tickle battles. Behavioral neuroscientist Jaak Panksepp at Bowling Green State University reports similar behavior in rats, complete with their ultrasonic (50 kHz) play vocalization, the rodent equivalent of laughter.[4] Panksepp's rats treat his tickling hand as a playmate, struggling, running away, but returning for more. Kenyan elephant expert Joyce Poole observed similar tickle bouts in piles of writhing pachyderms.[5] Fortunately, it was not necessary for her to dive into the pile to stimulate these cavorting beasts.

The labored breathing of mammals during such rough-and-tumble is the origin of human laughter, with the pant-pant of our primate ancestors evolving into the human ha-ha.[2] The unconsciously controlled panting laughter of our ancestors signaled that a physical encounter was play, not an assault. (Details about the emergence of human laughter and speech are provided in Chapter 2.) A lot of essay writing and speculation by humor theorists would be curtailed if their accounts started with this essential fact.

The tickle game begins during infancy and is a neurologically programmed means of communication between babies and mothers prior to the development of speech. Laughter signals "I like it; do it again!" Crying, fussing, and fending off the other person signal that the game has lost its appeal. Although tickling is probably a mammalian universal, only chimpanzee and perhaps other great ape (but not monkey) mothers and babies play the tickle game, with its associated mutual gaze and intimacies.[6] Feigned tickle of the "I'm going to get you" game is my candidate for the most ancient joke.[2] This playful ruse is the only joke that works well with both human babies and chimpanzees. The act of tickling does not qualify because the laughter produced is a reflexive response to the stimulus.

Christine Harris of the University of California, San Diego, demonstrated that the significant other providing the stimulus for vigorous tickling and laughter can be a "machine"—actually, a stooge who successfully masqueraded as a machine to the blindfolded experimental participants.[7]

The tactile friskiness and reciprocity of tickling develop into adolescent and adult sex play.[2] The evidence is striking. My questionnaire data revealed that from adolescence onward, you are about seven times more likely to tickle and be tickled by someone of the opposite sex than by someone of the same sex. The difference is even greater if you ask whom you would most like to have as a tickle partner. This pattern would probably be different among those who prefer same-sex partners. Such huge differences indicate the discovery of a fundamental process. (Conversely, the struggle to find tiny effects in huge samples may suggest that more care be taken in problem selection.)

Even confirmed tickle haters can conceive of situations in which being tickled might not be half bad. When asked when she most liked to be tickled, a twenty-three-year-old female noted on her questionnaire, "Never, really, but with a boyfriend in bed is OK." Her college classmate was more enthusiastic: "When my boyfriend tickles me, anytime, any place." Tickle was a popular form of sexual foreplay among both male and female respondents. "Tickling is the perfect way to touch girls with a good excuse," suggested a twenty-year-old male. "If you accidentally touch their private parts, you can always say, 'Oops, I'm sorry, I was just tickling you.'" This innovative tactic was not lost on a more demure twenty-two-year-old female, who noted, "The person may want to touch the other person romantically, but doesn't know how or when to start, so he/she tickles the other person to initiate the touching."

Erogenous zones are highly ticklish, although tickle scientists have neglected these areas, exploring instead the less controversial terrain of ribs, palm, and foot. Given the links between tickle and sex, it's not surprising that tickle fetish is a benign form of sadomasochism, in which torment is administered to restrained victims by fingertip and feather. Overlap is found between tickle and foot fetish websites (eg., In the Feet of the Night, Solefully Yours). In the discussion areas of such sites, I learned of the initial popularity and subsequent decline of foot stocks at Renaissance fairs. Fair managers were not amused by the discovery that tickle and foot fetishists may own and frequent such concessions. However, tickling and stocks really were a presence in the Middle Ages, and there are anecdotal reports of bloodless, sometimes fatal tickle torture, administered, in one colorful case, by goats licking salt from the soles of the victim's feet.[2]

The frequency of tickle frolics declines precipitously (about tenfold) after the age of forty, probably due to fewer potential ticklees after children have grown up and left the home, fewer potential sex partners, and a reduced sex drive. Older people also become less attractive tickle partners. When was the last time you desired to tickle or be tickled by an old person? Tickle's last hurrah may be in the nonsexual physical play of grandparents with grandchildren. Your recent tickle history is a reasonable predictor of your age, social, and sexual situation. When were you last tickled? And by whom?

Respondents of my study rated being tickled as moderately pleasant, with males enjoying the experience a bit more than females. For most, it was more blessed to give than to receive. On a 10-point pleasure scale (1, very unpleasant; 10, very pleasant), respondents rated tickling as a bit more enjoyable (5.9) than being tickled (5.0). (The midpoint on a 1-to-10 scale is 5.5.) "I like to tickle little babies on their feet because they laugh and squirm," reports a nineteen-year-old female enthusiast.

Tickle Me Elmo, a toy superhit now in a more expensive, high-tech second edition, is testimony to the pleasure of tickling. A press of the original Elmo's belly yields laughs and the comment, "That tickles." On the second press, laughs, followed with, "Oh, boy!" On the third press, simultaneous laughs and vibration, emulating a struggling ticklee. The ticklee's response is critical; it's no fun tickling an ordinary, immobile doll. Would anyone buy a Tickle Me Barbie that just lies there and does not struggle, laugh, or otherwise respond? Elmo's laugh, like that of human ticklees, triggers a contagious, unconsciously controlled laugh response in ticklers, and his vibration locks you into a primal, neurologically programmed bout of the tickle game. So far, there is no Tickle *You* Elmo that initiates the game by assaulting your tickle zones. As considered below, such tickle machines would be difficult to build, for interesting, non-trivial reasons, including the creation of a realistic animate entity.[2]

~

The sociality of tickle, a result of our inability to tickle ourselves, is the neurological basis of the simplest social scenario, *self* and *other*.[2,8] Self and other involve an intrinsic dualism—the emergence of self automatically brings other (nonself), what remains after the subtraction of self. This mechanism differs from the more familiar approaches to personhood of René Descartes ("I think, therefore, I am"), Martin Buber (*I and Thou*), or Carl Rogers (*On Becoming a Person*), offering an austere but straightforward alternative that applies equally well to nervous systems and machines. It is also a basis for a social neuroscience that can link the often estranged disciplines of personality, social psychology, and neurology.

The computation of self is the essential starting point of all things personal and social. The boundary of your personhood is mapped by your skin, the physical barrier that keeps your insides in and the outside out; the breaching of this boundary by alien entities such as bacteria and other antigens activates the immune system, another of your body's self/nonself discriminators. The itch/scratch system, described in the next chapter, also defends the skin and the self within. The perceived border of one's body is more nebulous than commonly thought, and sometimes subject to phantom phenomena when a body part is amputated, or to denial of ownership of part of one's body when the brain is damaged.[9]

The neurological mechanism computing self and other involves the cancellation of self-produced cutaneous (skin) stimuli. In the absence of such cancellation, we would be constantly goosing ourselves—lurching through life in a chain reaction of tactile false alarms. Pathology of this self/nonself (self/other) discriminator may play a role in atypical social behavior such as autism. Temple Grandin, in *Thinking in Pictures*, her fascinating account of life as a high-functioning person with autism, notes, "I always hated to be hugged," and "many . . . children crave pressure stimulation even though they cannot tolerate being touched."[10] Of particular interest here is that the aversion is not to touch itself but to the touch of others, a dilemma Grandin resolved by building a hug machine that provided her with the desired self-controlled touch without the unpleasant social (nonself) component. When touched by another person, does she experience something akin to your sensation of tickle, complete with the urge to fend away the stimulus and escape?

Might Grandin's aversion to social touch reflect hypersensitivity of her otherness detector, triggering an emergency response to stimuli that would be welcomed or unnoticed by most people? Conversely, might different social anomalies in some individuals be associated with hyposensitivity of their otherness detector, causing neglect of others, or a fuzzy self/other border? My test of the later hypothesis is limited to the examination of a single moderate-functioning autistic Down syndrome child who was, as predicted, insensitive to digitally administered tickle to the soles of her feet and ribs.

Given its ease and noninvasiveness, tickle tests should be administered to a wide spectrum of individuals with neurological and social anomalies. I predict that some individuals will show abnormal tickle responses associated with previously undetected classes of pathology. How, for example, do touch- and tickle-averse autistic individuals differ from autistic people who are not sensitive in that way? As considered in Chapter 1, contagious yawning, another nontraditional measure of sociality, has revealed novel deficits in the empathy of autistic children. Tickle studies may yield similar insights. Self/other detection may even have explanatory power in the general realm of personality. Are differences between "touchy-feely" and "prickly" people due to the fine-tuning of their self/other detector? At several levels, tickling is one of the deepest and most challenging problems in science. Even the determination of what makes a tactile stimulus ticklish involves the operational definition of self and other.

More than two thousand years ago, Aristotle observed that one "feels tickling by another person less if one knows beforehand that it is going to take place, and more if one does not," and "a man will therefore feel tickling least when he is causing it and knows that he is doing so."[11] Sarah-Jayne Blakemore and colleagues at the University of London extended and refined Aristotle's proposition, showing that the more unpredictable the stimulus, the more nonself and ticklish it becomes.[12] Using a robotic tickler controlled by a computer, Blakemore varied the correlation between joystick movements produced by a subject's left hand and a corresponding tactile stimulus applied by the robot to the subject's right hand. The subjects' ratings of "tickliness" increased with increasing delay (up to 1/5 second) and trajectory perturbations (up to 90°) between sinusoidal movements produced by the subjects' left hand and the sensory stimulation of the right. In other words, the stimulus becomes more tickly when it zigs when you zag. The light tickle (knismesis) evoked by the machine produces an urge to scratch or rub the site of stimulation or withdraw the stimulated part, not the more potent sensation that triggers squirming and hardy guffaws (gargalesis).

When you touch yourself, your brain compares the resulting stimulation to the expected stimulation and generates only a weak sensory response. When someone else touches you, the response to this unexpected stimulation is stronger. Apparently the brain gets messages about your own movements that subtract from the touch sensation. Using functional magnetic resonance imaging (fMRI), a noninvasive technology that provides images of both brain structure and activity, Blakemore implicated the cerebellum as the site where self-produced stimulation is cancelled. Only the cerebellum showed the selective lowering of neural responses to tactile self-stimulation relative to external stimulation, evidence of a cancellation process.

In my search for neurological insights based on everyday experience, a revelation about tickle and personhood appeared while I was soaping my foot in the shower.[2] I was surprised to find that my stroking the sole of my foot tickled more than it should have, given my understanding about one's inability to tickle oneself. The ticklishness was strongest when I tickled my left foot with my right hand or my right foot with my left hand, and least when I tickled my left foot with my left hand or my right foot with my right hand. This result was confirmed by twenty-five right-handed students in one of my classes (fully dressed and sans shower and soap).

They sat in a cross-legged manner in their chairs with the ankle of one leg resting on the knee of the other. They then strummed the naked sole of their foot with their finger tips, first with one hand and then with the other. The students rated contralateral (opposite-side) stimulation as more ticklish than ipsilateral (same-side) stimulation, with the most potent self-tickle occurring when the left foot was stimulated by the right hand. Relative to ipsilateral stimulation, our brain is less likely to recognize contralateral stimulation as self-produced and, therefore, generates a more intense sensation of tickle. The mechanism is probably associated with the greater disparity of arrival time of contralateral relative to ipsilateral information at the self/nonself comparator described by Blakemore.[12] The greater the arrival time difference, the greater the otherness, the less the selfness, and the greater the tickle. Take off your shoes and try this demonstration for yourself.

Does this tickle evidence suggest that one-half of our body treats the other half as relatively alien? I believe that it does. During the first year of life, body-left and body-right function in relative independence, slowly becoming coordinated as the commissural nerve fibers develop that couple the two sides.[13] Even in adulthood our two halves may coexist in harmony, but not perfect synchrony.

Providing pleasing symmetry to the chance encounter at the Goddard Space Flight Center that opened this chapter, I was recently invited to give another colloquium, this time to discuss the importance of building ticklish robots. I opened my presentation with the anecdote about tickle that occurred during my previous visit, but my anonymous muse, if present, did not come forward. Although the development of ticklish robots seems an unlikely priority, the tickle mechanism has practical implications for enhancing machine performance, just as it does for organisms. It is, after all, important for robots performing a fine-motor task to distinguish between touching and being touched by a manipulated object, and for roving robots to distinguish between objects they bump into and those that bump into them. Goosey robots, like goosey organisms, are unlikely to have fine motor skills, and will waste time and energy on tactile false alarms. Provocatively, the required machine algorithm for self/other discrimination may provide a computationally based construct for the emergence of body-image and machine personhood, complementing ongoing efforts by Cynthia Breazeal and Rosalind Picard at MIT, and others, to develop personable and emotionally responsive robots.

The problem of tickling is now handed off to readers, who, alone or with a willing partner—human or robot—can continue this research. But small-scale, no-budget science of this sort does have hidden costs. My wife watches me like a hawk for a certain gleam in my eye and other cues of impromptu, home-based tickle research.

11

itching and scratching

The itching of M.'s scalp was relentless and unbearable.[1] The nights were the worst, when she would scratch during sleep, waking in the morning with blood on her pillowcase. Hair was lost in the itchy area on her right forehead. Her internist was puzzled, and medicated creams didn't help. Scratching brought fleeting gratification but not lasting relief. The urgency of her problem became clear when M. awoke one morning with fluid running down her face. She placed gauze on her forehead and returned to her internist, who immediately called an ambulance and sent her to a hospital emergency room: during the night, she had scratched through her skull, all the way into her brain. Hospitalization for two years, during which time she slept with a foam helmet and hands tied to bedrails (later, with mittens on her hands), allowed the damaged tissue to heal, but the itching and urge to scratch persists to the present, though they are more manageable. She now lives at home but bears scars of her scratching, including partial paralysis on her left side. Although M. experienced an extreme case of pathological itch secondary to a shingles infection, we can all relate to her plight and understand why itching is a torment in Dante's *Inferno.*

Itching and scratching are linked, with itching (pruritus) defined as the unpleasant sensation that evokes scratching. Itching is a sensation of the skin, mucous membranes, and conjunctiva—it is never felt in muscle, joints, or internal organs. Good thing, because we could never get relief. But why do we itch and scratch? However pesky, the itch is adaptive, a useful prompt, like pain, that our body surface needs attention. Scratching whisks away some of our historical pests, including disease-carrying fleas, lice, ticks, flies and mosquitoes, as well as venomous insects and toxic plants. Itching joins tickling (Chapter 10) as part of the body's intruder alert and defense system. The skin is the body's first line of defense and must be kept in good condition. Itching becomes pathological only when it causes excessive scratching and tissue damage, which causes yet more itching, locking us into the seesawing itch-scratch cycle.

Scratching is so important for our well-being that it's a hair-trigger response to many cues, real and imagined, and prone to false alarms. Itching has many causes, including dry skin, parasites (fleas, lice), hives, scar formation, allergic reaction (poison ivy and poison oak), skin conditions (psoriasis, eczema, athlete's foot), thyroid disease, diabetes, Hodgkin's disease, jaundice, insect bites (mosquitoes, chiggers), drugs (opiates), dialysis, shingles, brain tumors, and multiple sclerosis. This is only a partial listing but indicates the great variety of itch-producing stimuli, some not directly related to the skin's surface. Itching can originate at any point in the sensory pathway between the skin and brain. The case of M., for example, was of brain origin because the viral attack of shingles that initiated her condition killed most sensory nerves from her itchy forehead, and attempts to block sensory input from her forehead with nerve-blocking agents and by cutting of sensory nerves had no lasting effect on her itchiness.

Itching persists in our relatively flea- and louse-free, hypoallergenic era, and is a leading complaint of patients who consult dermatologists. (Disfigurement and cosmetic concerns are others.) There is a ready market for over-the-counter itch remedies, including Lamisil (antifungal for athlete's foot) and the imaginatively named Anusol (hydrocortisone to reduce rectal swelling) and Vagisil (benzocaine for vaginal itch plus an antiseptic). Itching is so common that there is an unspoken scratch etiquette—if you must scratch, do so discreetly. Can home be defined as the place where you can scratch where it itches?

We are physiologically, psychologically, and socially primed to scratch. Simply viewing someone scratching may be enough to direct your attention to a previously neglected nether region of your own body that needs clawing (Fig. 11.1). I recall feeling itchy and uncomfortable during a meeting with a constantly scratching student who had an unsightly case of eczema. Simply thinking about itching while writing this chapter caused me to itch and scratch, and reading about it may have the same effect on you. It makes sense that itching and scratching are contagious; the pests bothering your neighbor may jump ship and infest you.

Itching and rash were certainly catching among students of a small rural elementary school in West Virginia in 1982.[2] At 9:15 a.m., two fourth-grade girls complained of itching and rash. Within minutes, other students were affected. By noon, thirty-two students had complained of similar symptoms, with fifty-two of 159 students in six classrooms affected by the end of day one. The itch crisis was covered in local newspaper and television reports, prompting parents to picket the school and discourage student attendance. The symptoms disappeared when the children left school and returned at the

figure 11.1 *Scratching a pesky itch can be done in solitude* (upper), *as a contagious response to seeing other people scratching* (lower left), *or by observing itch-related stimuli such as lice or skin with eczema* (lower right). *We are biologically primed to maintain the skin by engaging in the defensive and auto-grooming act of scratching. Our internal organs aren't itchy. Good thing!*

start of the next school day. Family members experienced no secondary infections. Public health authorities investigated the incident and concluded that the outbreak of itching and rash was not due to physical or infectious agents. By exclusion, a psychogenic cause was indicated, with the vector of contagion being word of mouth. The rash was apparently caused by scratching, because it was limited to areas accessible to the children's hands. Consistent with previous reports of hysterical behavior, young females were most affected. The symptom rate for girls was 51 percent and for boys 21 percent. The overall symptom rate among third- and fourth-grade students was 54 percent, and 16 percent among fifth- and sixth-grade students. The symptoms gradually abated over a two-to-three-week period, and no new cases were reported.

Volker Niemeier and colleagues at the University of Giessen in Germany pursued the issue of psychogenic itching in adults, finding that a lecture about itching, complete with itch-related images including mites, fleas, scratch marks on the skin, and allergic reactions, triggered scratching and increased ratings of itchiness by videotaped audience members (Fig. 11.1).[3,4] Another research group reported a spontaneous, stress-induced, psychologically mediated epidemic of itching and rash among elementary school children.

The visually mediated contagion of itching was explored more systematically by Alexandru Papoiu and colleagues at Wake Forest University.[5] Normal subjects and patients with atopic dermatitis (the most common form of eczema) received either itch-producing histamine or a saline control solution applied to the forearm, and viewed video clips of a person either scratching or sitting still. The sensation of itchiness and act of scratching were contrasted in the two subject populations. The scratching video had a much larger effect on the sensation of itchiness among eczema patients than among control subjects, even when the mock saline stimulus was delivered. The visually stimulated amplification of itching exists in healthy individuals but is much stronger in those with eczema. Eczema patients also scratched a lot more than healthy control subjects and its pattern was different: they were much more likely to experience itchiness and scratch far beyond the local itch induction site on the forearm, extending to the opposite side and distant parts of the body.

The potency of psychogenic itch stimuli, contagion, and the generalized pattern of response beyond the site of stimulation indicates the involvement of a higher brain mechanism. A challenge of treating itchiness may be that chronic itching sensitizes the brain to detect itching and increase scratching. Therapy may require both unlearning the itch-scratch syndrome and curing the local, underlying disease.

Pain and itching are antagonistic. Pain suppresses itching but is suppressed by opiates, and opiates that suppress pain generate itching. In a demonstration suggested by James Kalat of North Carolina State University, the pain/itch relation can be explored during a future visit to a dentist when having a tooth drilled.[6] When the Novocain starts to wear off, part of your still-numb face may feel itchy, anecdotal evidence that your itch system recovers before your pain system. Note that scratching your still-numb face may not relieve the itch because your pain system is still blocked; scratching cannot provide the pain necessary to block the itch. A lot can be learned from such informal self-experiments, including the design of future laboratory studies.

Given the great variety of itch-producing conditions and stimuli, it is not surprising that there may be at least two itch mechanisms.[7] One involves the release of itch-producing histamine by dilated blood vessels of mildly damaged, inflamed, or healing tissue. A second mechanism involves the response to irritant plants such as cowhage, the spicules of which are the insidious ingredient of itching powder. The two itch processes respond differently to treatment, and each may have its own neural pathways. Antihistamines block the itch produced by histamines, but not that produced by cowhage. Conversely, rubbing the skin with capsaicin, the source of heat in hot peppers, blocks the itch caused by cowhage, but not that of histamine.

The neuroscience of itching has gained momentum in recent years, with reports from the research frontiers having the exciting but unsettled quality of a work in progress. The story shifts from year to year, if not month to month. One of the major research themes is the paradoxical relation between itching and pain.[8] Itching and pain differ in the behavioral response to their stimuli—you approach and scratch an itch, but withdraw from pain. Itching is also inversely related to pain, being reduced by painful counterstimulation (e.g., scratching) and increased by opioid analgesics. Itching and pain have independent central pathways.[9] However, the similarities between itching and pain are so numerous that itching was long thought to be a weak form of pain: itch-producing agents can simultaneously cause both itching and pain; both pain and itch information are conducted from skin to spinal cord via small, slow-conducting, unmyelinated C-fibers; cutting a spinal cord sensory pathway (anterolateral funiculus) eliminates both pain and itching; and individuals with congenital insensitivity to pain are also insensitive to itching. Getting this story straightened out is a challenge but worth the effort. Chronic itching is a serious medical symptom, and solving the problem of itching may contribute to our understanding of pain.

In 1987, Hermann O. Handwerker and colleagues challenged the traditional itch/pain link by demonstrating that histamine-induced itching could be increased without increasing pain—different sensations involving different neurological pathways must be involved.[10] A decade later, Martin Schmelz and colleagues discovered sensory receptors in the skin for itching.[11] By injecting tiny amounts of histamine into the skin and recording its effect on sensory nerves, they found receptors that responded exclusively to histamine and produced a corresponding sensation of itching in experimental subjects. Each sensory nerve for itching collects information from a very large area of skin (8.5 cm dia.) versus the much smaller catchment area (2.4 cm dia.) for painful stimuli.

The collection area for each sensory nerve (its receptive field) is negatively correlated with the spatial resolution for the sense, with smaller fields providing better resolution. This is why it's difficult to precisely locate the source of an itch; you can't tell where within the broad catchment area the stimulus originates. The peripheral nerve fibers conducting information for both itching and pain are very slow (0.5 and 1 meter per second, respectively), about a hundred times slower than the sensory superhighway transmitting information about light touch. This is the reason why you can sense something touching your skin before signals arrive making it itchy or painful. The C-fibers that mediate itching can be activated directly by skin irritants or indirectly by immune system cells that monitor the skin and release histamine when they encounter foreign substances such as mosquito saliva.

Steve Davidson and colleagues at the University of Minnesota identified neurons that are active during itching and calmed by scratching.[12] The neurons are in the spinothalamic tract of the spinal cord, which receives itch signals from sensory nerve fibers in the skin and relays the signals to the brain. To see if scratching could calm the spinothalamic neurons, Davidson injected the legs of sedated monkeys with histamine, a known itch-producing substance. The histamine wildly excited the spinal neurons, and the experimenters scratching the leg with an imitation monkey hand (the monkey was sedated and couldn't scratch itself) reduced their firing, but only during an itch. These data suggest that the scratch-related cutoff switch for the itch sensation may be in the spinal cord, but the story is still unfolding. An important implication of these findings is that all types of itch, including those not involving histamine, could be treated by drugs that mimic the inhibitory effect of scratching. A pill could do your scratching for you, without the risk of tissue damage.

Zhou-Feng Chen and colleagues at Washington University (St. Louis) discovered the first gene related to itchiness, named GRPR (gastrin-releasing peptide receptor).[13] Mice with an inactive version of this gene scratched less when exposed to itchy stimuli than mice with an active gene. To prove that the neurons that had this gene were itch-specific, Chen's team injected the spinal cords of mice with a neurotoxin that destroyed most of the cells harboring that gene.[14] The injection dramatically reduced and sometimes eliminated scratching caused by itchy stimuli of all types. The treated mice were not simply numbed, because they could move normally and respond to other stimuli such as pain. This is the most convincing evidence to date for an exclusive "labeled line" for the transmission of the itching sensation and the independence of itch and pain information.

Ants marching across your foot or a spider crawling up your neck are evidence that mechanical stimuli can immediately trigger a scratch-like defensive movement that does not involve the just-considered allergens, histamine release, immune system, chronic pathology, or slowly conducting C-fibers that have attracted dermatologists and other researchers. The titillating sensations of light touch are more tickly than itchy, but they join itching in initiating defense of the body surface. Both tickling and itching are complementary self-limiting sensations, triggering actions to terminate the source of simulation. Unlike itching, the sensation of being tickled is less noxious, probably involves fast-conducting large-diameter myelinated fibers, and has a stronger social component (because you can't tickle yourself). The animate other, the "not you" stimulus necessary for tickling, can be provided by another person or by the just-mentioned ants and spiders. The response to a tickle is a relatively variable whisking away of the invading entity or offending hand, not the more stereotyped act of scratching. Itching lacks "thingness"—the animate "other" defining the tickle stimulus.

~

Scratching is a neglected part of the itch story, although it is central to operationally defining itching, the sensation that makes us scratch. Scratching deserves more respect. Itching, after all, has no adaptive significance without the scratch that quenches it, but scratching is functional without the itch. Spontaneous scratching grooms the body surface, removing debris, parasites and irritants, but of what use is itching without scratching—a diabolical torment that can't be relieved? Scratching must have evolved before itching, the sensation that guides a scratching appendage to itchy places. Likewise, the sensation of pain must have evolved after the motor act needed to stop the noxious stimulus. Of what use is unrequited suffering?

Scratch research gets lost in a research area whose understandable priority is getting the damn itch to stop. But scratch researchers reciprocate by neglecting the itch; there is a surprising insularity of the respective literatures. Although nonprofessionals may consider scratching and itching to be essentially the same thing, researchers have spent careers studying one or the other topic without integrating them or citing key papers in each other's specialty. Curious.

The science of scratching has a distinguished history, including exemplary work done more than a century ago by Sir Charles Sherrington, Nobel laureate and patriarch of neurophysiology.[15] From Sherrington to the present, researchers appreciated that the scratch reflex was an ideal approach to fundamental neurological problems, especially those involving how movement is produced and controlled. Neurological mechanisms are most easily understood when the movement that they produce is simple in form, stereotyped (repetitive), and easy to observe and measure. Scratching is such a movement.

Whether in rodent, dog, or cat, we have witnessed the rapid, stereotyped movements of a limb scratching a body part to attack a flea, real or imagined. Sensory input can be eliminated as the source of the repetitive scratching, because such patterning does not exist in the stimulus; a single flea bite or mild shock from an electric flea triggers a flurry of scratching, not one scratch cycle per bite or shock.

This rapid-fire sequence of limb movement must be programmed by the spinal cord, not the brain, because there is insufficient time for one scratch cycle to trigger the next—nerve impulses must make a round trip to the brain and back for that to happen. A spinal origin is confirmed because scratching remains in animals that have had their spinal cord transected, isolating it from the brain.[14] In fact, it's easier to evoke a scratch response in such animals than in intact animals, perhaps due to the release of the cord circuitry from brain inhibition. The repetitive scratch pattern also remains after sensory nerves from the scratching limb to the spinal cord have been severed, eliminating another possible source of sensory feedback that could influence the motor pattern. By elimination, the form and timing of scratching must be orchestrated by a so-called oscillator or motor pattern generator within the spinal cord. Sensory input, perhaps a flea bite, turns on the neurological oscillator, but it does not pattern its motor output.

If this historical focus on mere scratching seems extravagant, recall the principle of biological conservation mentioned in the book's introduction. Once evolved, neurobehavioral mechanisms are adapted for many purposes. Oscillators similar to that of scratching also produce the rhythmic movements of walking, flying, and swimming. As with scratching, the brain does not micromanage the neuromuscular details of movement, but simply commands "walk," "flap," "swim," or "scratch," and the spinal cord oscillator executes the order, sending out detailed information about when specific muscles are to contract. Sensory input may initiate or modulate the intensity or duration of the behavior, but it does not produce its pattern.

A fascinating property of the scratch reflex is that the limb is directed toward the site of stimulation, even in animals who have had the brain disconnected from the spinal cord. Such animals are responding to an itch that is unknown to their brain. Does this mean that the isolated spinal cord has a primitive form of awareness, or that awareness is irrelevant to understanding an automatic response?

Unlike the frequently studied dogs, cats, turtles, and frogs, humans with a complete transection of the spinal cord may not perform a scratch reflex. The isolated human spinal cord may lack sufficient excitability to produce the reflex. But humans deep in coma who are not responsive to strong visual or auditory stimulation sometimes respond to firm pressure to the chest by producing arm movement, the site-specificity of which is a positive score on a neurological exam for comatose patients.[16] During lowered states of awareness, the ancient scratch reflex may be active in individuals with intact nervous systems, as suggested by the case of M. at the beginning of this chapter, whose most damaging scratching occurred during sleep. Sleep may release (disinhibit) the primal circuitry for scratch.

~

Grooming is a slower, more deliberate, and more systematic means of maintaining the body surface than the reflexive, stimulus-driven bursts of scratch and tickle. Think of a grooming animal as a biological machine that provides its own maintenance, like a car that periodically washes and waxes itself. The great antiquity and value of grooming are indicated by its performance by a huge variety of animals, from insects to humans, and its mediation by unmyelinated nerve fibers. Individual (auto) grooming can be as varied as a cat preening its fur, a termite cleaning its antennae, or a lobster removing debris from its leg. Animals can also groom each other. Social insects such as termites, ants, and bees engage in mutual (allo) grooming, contributing to individual and group fitness. Our fellow primates move this hygienic practice into the political arena, using social grooming to establish and maintain alliances and dominance hierarchies.

Grooming, whether personal or social, is a measure of the fitness of the individual and group. A reduction in personal or social grooming, whether naturally occurring or experimentally induced, predicts pathology, declining health, and, sometimes, impending death. That the three complementary behaviors of scratching, tickling, and grooming evolved independently to maintain the body surface is a measure of the importance of the function. The significance of the process is reflected in its low threshold. Even thinking or reading about itching may get scratching under way.

12

farting and belching

Some people possess gifts so extraordinary that they define what is possible for our species. Such a man was Frenchman Joseph Pujol (1857–1945), stage name "Le Pétomane."[1] Pujol's unique gift was based on his ability to "inhale" through his rectum, an act accomplished by expanding his abdomen while ceasing to breathe through his nose and mouth. Air thus inhaled could be exhaled with force sufficient to extinguish a candle from a distance of a foot—no mean feat. But such tricks were neither the true measure of his artistry nor, as we will learn, his unappreciated contribution to science.

Pujol's breakthrough came as a child playing on the beach when he was hit by a wave of cold water. He was startled and somehow sucked sea water into his rectum. Young Joseph, being a curious lad of experimental bent, soon discovered that he could inhale air as well as seawater though his rectum, and could exhale this air in a controlled manner. Unlike lesser men, Pujol was an entrepreneur who appreciated the comic and musical potential of a controllable anal air flow that lasted ten to fifteen seconds. He honed his craft in the provinces before setting out for Paris and auditioning for an astonished Charles Zidler, the manager of the Moulin Rouge, the world's most famous nightclub.

Pujol was a direct sort who believed in immediately getting down to business in his interview. After a traditional introduction, he dropped his trousers, dipped his bottom into a large washbowl in Zidler's office, and sucked the water into his rectum, then promptly refilled the bowl to cleanse his instrument. This is the kind of job interview that's likely to be a deal maker or deal breaker. In Pujol's case, it succeeded wonderfully and led to one of the most remarkable careers in show business. History does not record whether Pujol's interview involved another of his feats, expelling water from his rectum in a mighty jet that could reach twelve to sixteen feet, but he did demonstrate the vocal range of his instrument, from tenor to baritone. We will follow Joseph Pujol's career under his stage name, Le Pétomane—literally, "fartomaniac."

Le Pétomane's act was full of novelty, comedy, and virtuosity (Fig. 12.1). At his peak, Le Pétomane easily outearned the great actress Sarah Bernhardt, his closest contender. And what was the act of this artist who "pays no author's royalties"? He would begin with a series of ordinary farts, describing each in turn—a little girl, a mother-in-law, a bride on her wedding night (weak) and on the morning after (loud), tearing cloth, cannon fire, and thunder. With a tube inserted in his rectum, he would smoke a cigarette or attach a flute and play tunes. But his real artistry was accomplished au naturel.

Le Pétomane's repertoire included animal sounds—a rooster crowing, a puppy, a dog with its tail caught in a door, a blackbird, an owl, a duck, bees, a tomcat, a toad, a pig, and musical instruments including violin, bass, and trombone. The climax of Le Pétomane's performance was a stirring rendition of "La Marseillaise" that brought down the house. He

figure 12.1 *Le Pétomane ("Fartomaniac") was the main attraction at the Moulin Rouge from 1892 to 1914. His repertoire was impressive, ranging from hilarious sound effects to a rousing rendition of "La Marseillaise." His technique involved the use of his rectum as a bellows, sucking air in and expelling it through his anus. Buttspeak was never a serious contender in human speech evolution, with the vocal advantage going to the more flexible and somewhat more elegant laryngeal system.*

made audiences delirious with joy, writhing in their seats, tears streaming, with some fashionably dressed and corseted women so overcome that they were carried into hallways to be revived. Amazingly, no part of his performance was faked, as he once demonstrated in the nude before a panel of earnest and curious physicians.

Sadly, we cannot experience Le Pétomane's artistry because recordings of his act are unavailable, although one wax cylinder recording is rumored to exist. And he was born too late to win a fan in Mozart, who thought long and hard about flatulence, and who might well have been inspired to compose a concerto for Le Pétomane's very special instrument.[2] Certainly, Saint Augustine (AD 354–430) would have been impressed by Le Pétomane. In his *City of God* (Book XIV, Chapter 24), this pious advocate of willpower claimed knowledge of someone with such control of his rectum that he could break wind continuously at will and produce the effect of singing. To Augustine, Le Pétomane would have been a spiritual revelation. To those of us with more scientific interests, we are left wondering how Le Pétomane would have fared if he had turned his virtuosity to buttspeak—speaking through his rectum. Granted, buttspeak does seem a very dark horse in the vocal evolutionary sweepstakes, but looking at the possibilities of this option of speech origin is informative.

~

My foray into fart science is a bit timid. The mere inclusion of the topic threatens to lower this book's intellectual tone. A confluence of circumstances forced the subject on me. While recording laughter for an earlier study, one of my subjects laughed so hard that he farted. Since I already had it on tape and was in a sound lab, why not check out the acoustics of farting? This was a defining moment. Would I lose resolve, as did Galileo when he was "shown the instruments" by his inquisitors? With tenure safely in hand, I forged ahead. What started as a playful acoustic analysis led to the quite serious consideration of why we speak through the mouth instead of the rectum. Along the way, I discovered a quirky and amusing literature that may elevate the status of the lowly fart as a topic in scientific discourse.

Throughout the ages, farting (flatulation) has generated jokes, folklore, etiquette, and a few legal sanctions, but little research. The legendary Hippocrates (460–377 BCE) considered the medical affliction of too much gas in *The Winds*, advising, "It is better to let it pass with noise than to be intercepted and accumulated internally" (1.24–25)[3]. The topic has been treated more often in popular fare, including the humorous writings of Geoffrey Chaucer, Benjamin Franklin, and Mark Twain. Fart jokes earn a place in the opening scene of Aristophanes' *The Clouds* and the memorable closing scene of "The Miller's Tale" in Chaucer's *The Canterbury Tales*. Although passing gas is usually considered ill-mannered, it's usually laughed off as a minor offense. But this has not always been the case. The Roman Empire once had laws against farting in public places, a sanction that must have caused a lot of fanning and finger-pointing in the Forum. The law was lifted during the reign of Claudius, one of the most flatulent of emperors.[4]

Consider the sad fate of Pu Sao of the Tikopia in Polynesia, who was so overcome with shame after farting in the presence of the chief that he committed suicide by climbing a palm tree and impaling himself through the rectum with a sharply pointed branch.[5] Sanctions are less severe among the Chagga of Tanzania, but feminists have a lot of work to do there. If a husband breaks wind, the wife must pretend that it was really she who discharged, and she must submit to scolding about it. Failure to accept responsibility can cost the negligent wife three barrels of beer.[6] Unlike the Chagga, modern Americans are pretty much adrift regarding farting etiquette, with manners mavens such as Judith Martin (aka "Miss Manners") offering no guidance.

Although farting etiquette is varied, the medical establishment generally supports Hippocrates' position that it is better to let gas out than hold it in. Almost two thousand years later, the always surprising and encyclopedic Michel Eyquem de Montaigne in his essay "Of the Force of Imagination" notes forbiddingly that "God alone knows how many times our bellies, by the refusal of one single fart, have brought us to the door of an agonizing death."[7] Indeed! Modern medicine offers some support, reporting that flatus retention is the major factor in diverticular disease.[8]

The origin of the gas that powers farting has been the subject of speculation since ancient times. Ancient encyclopedist Pliny the Elder (AD 23–79) claimed that lettuce breaks up flatulence but garlic, leeks, and onions cause it, noting that Democritus (460–370 BC) totally opposed turnips as food because they cause flatulence.[9] Two millennia later, we have entered the golden age of flatulence research. Analytical chemists tell us that at least 99 percent of bowel gas is composed of nitrogen, oxygen, carbon dioxide, hydrogen, and methane (swamp gas), with all but oxygen and nitrogen arising from processes within the gut.[4] None of these flatus gases can be detected by smell; the odiferous component must come from other trace gases produced by gut bacteria.

Methane is the gas that sustains the odd male ritual of fart lighting, the ignition of farts by a match placed near the anus.[4] The resulting methane jet produces a lovely *blue* flame, quite unlike the orange flame portrayed in Jim Carrey's film *Dumb and Dumber*, the flatus-fueled conflagration in *Dennis the Menace*, and more recently that seen in Eddie Murphy's *Nutty Professor II: The Klumps*. Fortunately, only about one-third of the population can generate combustible levels of methane. On a more serious note, a gassy gut can be fatal, as it was for a patient having a colonic polyp cauterized. An electric spark caused the patient's bowels to detonate, blasting out the colonoscope and ripping a six-inch hole in the patient's large intestine.[10] Le Pétomane's farts would not have been a fire hazard—they were composed solely of noncombustible normal air sucked into his rectum.

~

Farting is an acoustic as well as chemical event, and I wanted to subject it to the type of analysis used previously for laughs, coughs, sneezes, and hiccups. It should be easy to collect fart sounds for analysis; the average twenty- to thirty-year-old farts about thirteen times per day.[4] However, not wishing to pursue the indelicate task of collecting further data, I will confine my analysis to the already collected sample from the laughing subject.

The acoustic structure of a fart is shown in Fig. 12.2. The bottom trace is a sound spectrum that shows strong harmonic structure, as reflected in the regular stack of frequency bands that are multiples of a fundamental frequency of around 150 Hz at mid-burst. The fart has a tonal quality and is clearly not a noisy blast with a random distribution of frequencies. The fart also has a periodic, pulsatile quality (amplitude modulation) illustrated by the regularly spaced vertical bands of the middle spectrum, and the waveform at the top of Fig. 12.2. A raspberry or Bronx cheer, produced by placing the tongue between the lips and blowing, makes a similar sound. Farts lack the structural stereotypy of laughs, coughs, sneezes, and hiccups, and their duration is determined by the highly variable supply of available gas. The artistry of Le Pétomane is testimony to the flexibility of this channel of communication.

We now rejoin the evaluation of buttspeak as communication medium. This question is not as outlandish as it seems and involves central issues in vocal production and evolution. Before you prematurely reject the possibility, consider the evidence. The traditional vocal apparatus of humans is not a paragon of biological elegance. No part of it evolved exclusively in the service of sound making. We speak through the same toothy orifice through which we breathe, eat, drink, and vomit, and the vocal folds (cords) used for sound production are simply two flaps of muscular tissue that act as a seal to keep food and drink out of the airway when swallowing. We choke when failing to operate this clunky, complicated, and delicate apparatus correctly. Sometimes it malfunctions on its own.

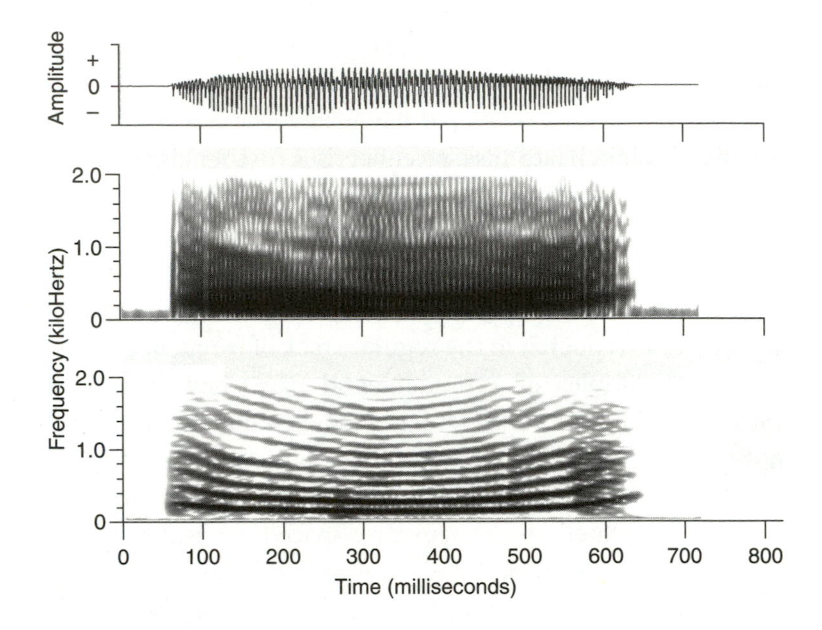

figure 12.2 *A waveform* (upper) *and two frequency spectra* (middle, lower) *of a fart of a young, adult male. Farts have a pulsatile, buzzing quality, as reflected in the discrete vertical bands in the waveform* (upper) *and spectrum* (middle). *Using a different filter setting, the lower spectrum reveals the striking harmonic structure of the fart, as reflected in the stack of horizontal frequency bands that are multiples of a fundamental frequency, the lowest band. Both farts and belches have highly variable durations that are dependent on the gas available for the egress (exhalation).*

The lungs and muscles of the thorax are bellows that propel air through the vocal folds, which respond by vibrating, producing a buzzing sound like a reed in a woodwind instrument. The upper vocal tract (throat, mouth, tongue, teeth, lips) provides resonance chambers and a means of shaping the buzzing sounds of the vocal folds. We "play" this biological instrument to produce speech.

Given the curious heritage of our vocal apparatus, butt-speak does not seem so far-fetched. If you use the abdomen and colon for the air reserve and bellows instead of the lungs and thorax, and the seal of the anal sphincter as the vibrating element instead of that of the vocal folds, you have a start toward a viable speech generator. With apologies to Louis Armstrong and Wynton Marsalis, the anal sphincter functions like the lips of a trumpet player buzzing into a mouthpiece, but in farting, the rest of the trumpet, the part that shapes the sound, is missing. (An adventurous musical acquaintance acknowledged that a tone can indeed be produced by farting into a trumpet, creating a middle C, but further details were not forthcoming from this wary pioneer of the butt-trumpet.) The traditional vocal tract offers much more flexibility to shape and tune an utterance produced by a vibrating sound source than the less traditional one available to Le Pétomane. He could honk out the tune of "La Marseillaise" but not its rousing lyrics.

Having made the case that farting could provide a novel though limited channel of communication, we can question whether any animals actually exploit this unlikely aural niche. So far, signaling by fart is reported only in certain herring. The breakthrough research appears in a paper cryptically titled "Pacific and Atlantic Herring Produce Burst Pulse Sounds."[11] The work earned Ben Wilson and colleagues at the Bamfield Marine Science Centre in British Columbia, Canada, the 2004 Ig Nobel Prize in biology. The scientists became curious about the origin of strange rasping noises produced at night by captive herring under observation in a tank, and discovered that they were fish farts. The fish gulp air at the surface, store it in their swim bladder, and release it from a duct in their anus, producing a high-frequency burst up to 22,000 Hz. (Human hearing ranges from 20 Hz to 20,000 Hz.) Their next insight was that the farts are signals that may bring fish together and assist in evading predators. But signaling may have a price: predators such as killer whales may also be listening to flatulent herring and home in on the signal. The herring fart by night, but not by day, when they rely on visual instead of auditory information.

~

Belching, like farting, is another potential path not taken in the course of human speech evolution. The sound of belching (eructation) is caused by the vibration of the upper esophageal sphincter as gas passes through it. This sphincter reflexively opens during swallowing, then closes, providing a seal against backflow from the esophagus into the pharynx. A lot of sonic energy can be produced by a belch—the current Guinness World Record is 118 decibels, set by Paul Hunn of London, England, in 2000.[12] This is louder than a chain saw at a distance of one meter.

The waveforms and frequency spectra of three typical belches are shown in Fig. 12.3. Like the fart, belches have a strong pulsatile character as reflected in the vertical bands (amplitude modulation) of the waveform (upper) and frequency spectrum (lower), but the pulsations of the belches have a much lower frequency. The noisy sonic blast of the belch is less tonal and lacks the marked harmonic structure of the fart (Fig. 12.2). Like the fart, the duration of a belch is limited by available gas.

Belching is usually caused by eating or drinking too fast and thereby swallowing air, which is subsequently expelled, causing the characteristic sound. Belches are also caused by imbibing carbonated beverages such as soft drinks or beer, in which case the expelled gas is carbon dioxide from the drink itself. Expelled gas can arise from digestive processes in the stomach or from gastroesophageal reflux. Babies are particu-

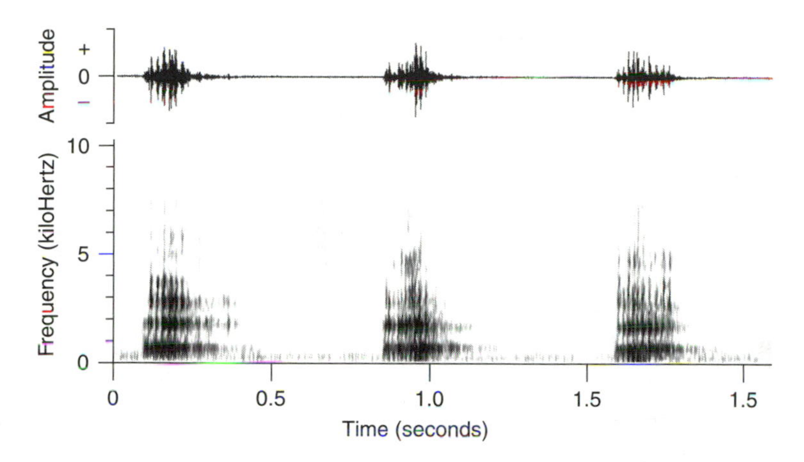

figure 12.3 *Waveforms* (upper) *and frequency spectra* (lower) *of three belches of an adult male that were facilitated by drinking a carbonated beverage. Belches, like farts, have a strong pulsatile quality, as reflected in the vertical lines in the waveform and frequency spectra. However, the spectra show belches to be noisier, broadband blasts that lack the strong harmonic structure of the fart. Three belches are placed on the same axis for purpose of comparison: the intervals between belches do not reflect actual intervals between belches.*

larly prone to accumulate gas in the stomach while feeding, causing distress until the baby is burped, releasing the gas.

Most human belches are composed of nitrogen and oxygen, the main components of swallowed air. A grazing cow, in contrast, releases up to six hundred liters of methane into the atmosphere per day. Contrary to folklore, 95 percent of this release is through burping, not flatulence. Researchers at the Commonwealth Science and Industrial Research Organization of Perth, Australia, are seeking to reduce release of this greenhouse gas by developing an anti-methane vaccine to minimize burping and flatulence in cattle and sheep.

Ordinary, belching people can achieve a level of virtuosity that puts Le Pétomane to shame. Belched (esophageal) speech can be produced voluntarily by swallowing air and then expelling it, manipulating the sound with the vocal tract as you would with normal speech. Children sometimes entertain themselves with this vocal novelty, but it is put to serious use by people who have experienced laryngotomy. With practice, esophageal speech can be used to produce intelligible words at a rate of around five words per breath, or about 120 words per minute, in a characteristically low pitch (50–100 Hz). A less demanding way of producing the vibrations necessary for non-laryngeal speech is the electrolarynx, a small device that is held against the throat. You may have heard the robotic, machine-like quality of speech produced by people using this device.

A theme of this chapter is how sounds are produced by vibrating structures and subsequently selected and shaped for auditory communication. This is an analysis of speech evolution in disguise. The necessary vibration may come from the upper esophageal sphincter in belching, the anal sphincter in farting, or the vocal folds, the structure ultimately selected for speech. All are elastic seals set in motion when they can no longer resist the passage of gas that they restrict. Belching, farting, and speaking all use bellows to provide the pressurized air supply necessary for vibration—the thoracic cavity in belching and speaking, the rectum in farting. With human beings, whether the vibrations are produced by vocal folds, the esophagus, or an electrolarynx, the vocal tract (throat, mouth, tongue, teeth) is still used to modulate the resulting sound to create intelligible words.

13

prenatal behavior

You are tumbling, submerged and weightless, in a warm and watery place, breathless but not suffocating. Your body jerks, twists, and bends, grasped by seizures out of your control, restrained by an unseen tether. The wet darkness is punctuated by pounding heartbeats, your own and another of unknown origin. To this relentless cardiac cadence are added occasional rumblings, gurgles, and periodic weak, higher-pitched bursts of what you later know to be speech. Sound terrifying? It's part of the universal human experience, although you don't remember it. Such is prenatal life, a period brought to a dramatic close by a long, difficult squeeze through a too-narrow orifice, and expulsion into the bright, cold, noisy world where we spend the rest of our days. Scientists have an equally challenging time intellectually traversing the birth canal in the retrograde direction, trying to understand the prenatal experience. Embryos are perversely indifferent to our efforts to understand their alien realm. Most psychologists, physicians, and other students of human development are largely unaware of what embryos do, why they do it, and why it's important. (While there is traditionally a distinction made between embryos and fetuses, based on weeks of gestation and stage of physical development, I will use "embryo" as a general term here to emphasize the continuity of the developmental process.)

Little that we know about life after birth prepares us to understand the period before we are born. The prenatal and postnatal forms of organisms, or even embryos at different stages, are so different in structure and function that if it were not for the thread of developmental continuity, they would seem members of different species. Piaget and Freud taught us that children are not miniature and incompetent adults, but intellectually and psychosexually distinctive beings who are pursuing their own stage-specific tasks. A comparable lesson is needed for prenatal life. Embryos, as we will see, are profoundly un-psychological beings with their own agenda. They are not simply small, wet babies incubating in a warm, dark womb, and their study is not well served by the research priorities of postnatal life. Even the structure of scientific and professional disciplines complicates our developmental mission. Developmental psychology, for example, usually does not begin its study of individuals until after birth, and it underemphasizes biological factors. Embryology errs in the opposite direction, arbitrarily ceasing study at birth and neglecting behavior. In medicine, the developmentally unimportant event of birthing casts a baby from the obstetrics department over to pediatrics.

Redress of the long neglect of prenatal behavior is a tall order for a short chapter. A more manageable task is to establish an *embryocentric perspective*, a framework for thinking about early development, and to provide a sample of some of the extraordinary phenomena of prenatal life.[1,2]

~

The most critical and least understood events in human life occur between conception and birth. We begin as a maternal parasite evolved to survive in the ecological niche of the womb, and are equipped with highly specialized adaptations to that alien, aquatic domain. The placenta, for example, is developed by the embryo, not the mother, as the biological link with its maternal host. In nine months, we go from being a fertilized egg (the zygote) to a vast constellation of cells called a baby, a complete human being ready to shed its lifeline and fend (with parental assistance) in the postnatal world.[3]

Our exploration of prenatal behavior begins with a brief review of how embryos work and what makes them special. These insights are the hard-won product of a research program starting with Aristotle, experiencing breakthroughs during the golden age of experimental embryology in the first half of the twentieth century, and continuing into the modern era of molecular biology and genetics.

Cells of the embryo arise through proliferation, the mitotic division of the zygote. The daughter cells of this division differentiate into specific cell types and functions, and increase in size. These processes of proliferation, differentiation, and growth (maturation) are the basic mechanisms of embryonic development. The zygote first is transformed into a tiny hollow ball of cells and then becomes an embryo, a step toward a recognizable organism. During development, cells are born, and many die; some migrate like schools of fish (or amoebas) through the tissue matrix of the embryo to reach their final position; and vast sheets of cells fold inward or outward as they differentiate into their mature form. This choreography of development is complex and fraught with risk.

Much can and does go wrong. Fortunately for us, the developmental process has intrinsic error correction for most such mistakes. Without it we would be monstrous creatures, if we were even able to survive the developmental process. Teratology, the study of such gross developmental errors, is a subdiscipline of embryology.

The paradox of a developmental process that is both imprecise and has a high success rate is resolved by classic discoveries of experimental embryology. Consider, for example, the *totipotency* (developmental flexibility) of the two daughter cells of the first cell division. If experimentally separated, each cell will form a complete embryo, not a half embryo. Thus, the fate of the daughter cells is not predetermined at the time of their birth but governed by an *epigenetic* process through which structure gradually emerges, influenced by the cells' milieu. Later in development, areas of the embryo differentiate into specific structures such as limbs. Again, the fate of cells is not strictly predetermined. The tissue that forms a limb is part of a *morphogenetic field*, defined as a region able to recover from the effects of partial ablation (destruction). In other words, if half of the limb's morphogenetic field is experimentally destroyed, the embryo will still develop a reasonable approximation of a complete limb, not a half limb. The more complete the field, the better the outcome. The implications of this result are profound.

A morphogenetic field has the property of *self-organization*, forming the best possible whole from available cells. A cell removed from its neighborhood and placed in tissue culture lacks direction and will not differentiate. Morphogenetic fields will not work if the fates of cells are predetermined, because they lack the requisite plasticity. Cells within a morphogenetic field are coordinated by a chemical gradient that operates by diffusion over short distances, the reason why all embryos are small and about the same size, whether mouse or great blue whale. Big embryos won't work because distances are too great and critical intercellular communication would break down.[4]

Consider the error-correcting properties inherent in a morphogenetic field. If a cell in the field is missing, another will be reprogrammed to take its place; if an errant cell wanders into the field, its developmental program will be overridden by that of its neighbors, in both cases forming the best possible whole. In such a self-organizing, cellular ecosystem, the loss of a cell or cells need not be catastrophic. These probabilistic, epigenetic processes force a reconsideration of what it means to be "genetically determined" and illustrate why genes are better understood as recipes than blueprints. Genes provide instructions for assembly, not a detailed plan for the finished product. This conclusion may be a revelation to individuals who think embryos and their behavior an uninteresting playing out of a genetic program. Embryonic cells are, by necessity, highly social entities in constant communication, shaping and being shaped both by each other and by more distant biological cues.

Self-organization also exists at the behavioral level. The examples explored in this chapter—the formation of ball-and-socket joints and the regulation of motor neuron numbers—were selected both for their significance and for their neglect by students of behavior development. We will return to these phenomena after considering the equally remarkable and unappreciated mechanism producing prenatal behavior.

~

Unlike most postnatal behavior, embryonic movement is neither goal-directed nor shaped by its consequences, at least not the kind considered by B. F. Skinner and his behaviorist descendants. Embryos are training neither for the Olympics nor for the SATs, but have more immediate and unanticipated developmental tasks. Embryos should be consulted for guidance in these matters because both their behavior and the mechanism that produces it are unique to prenatal life.[1,5,6] Embryos don't behave like small, klutzy adults—reaching, grasping, or stepping—who gradually, through growth and practice, become strong, skilled, and graceful. Instead, embryos twist, curl, twitch, and jerk, with new body parts becoming involved as they differentiate and become capable of moving. Recognizable acts such as yawning, stretching, hiccupping, and even thumb sucking occur early on, but they are not the norm. Likewise, sensory responses, habituation, and conditioning occur, certainly at later stages, but such behaviors are more significant for understanding postnatal life than for insights into prenatal life. There are better explanations for what embryos do and why they do it.

Quickening, the onset of fetal movement detected by the mother, usually begins around the eighteenth to twentieth week of gestation. (Full-term birth usually occurs thirty-eight weeks after fertilization, or forty weeks from onset of the last menstrual period.) Although quickening is a significant event for all pregnant women—how could they ignore it?—it's not biologically significant. Behavior starts much earlier, unfelt by the maternal host.

Johanna de Vries, now at the University of Amsterdam, with developmental neurologist Heinz Prechtl, pioneered the use of noninvasive ultrasonography to observe the development of human prenatal behavior.[5,6] Ultrasonography is a huge technical advance. Previous evidence from a century or more earlier was based on unsystematic maternal reports, attempts by physicians to discern in utero events via touch and stethoscope, and incidental observations of dying, surgically removed, or aborted fetuses.

De Vries observed the first spontaneous movements of normal human embryos around seven and a half weeks postmenstrual age, almost three months before they can be detected by the mother. During the following weeks, these feeble, early movements are joined by embryo-specific startles and unorganized general movements, as well as more familiar acts such as hiccups, yawns, and breathing. Although their relative frequencies vary, all of de Vries' sixteen categories of movement are achieved by fifteen weeks, more than ten weeks before babies born prematurely can survive.

Ultrasound yields invaluable descriptions of natural, ongoing prenatal human behavior, but experimental studies of animals provide the best understanding of its origin. Over a century ago, William Preyer showed the way in his monumental book *Specielle Physiologie des Embryo* (Special Physiology of the Embryo, 1885), a companion to his better-known *Die Seele des Kindes* (The Mind of the Child, 1882).[7] Although not the first person to observe the behavior of embryos and fetuses, his comparative and interdisciplinary work was the most comprehensive and original. Among Preyer's discoveries was *motor primacy*—that embryos "spond before they respond" (act before they react), and that their behavior is produced by spontaneous activity within the nervous system, especially the spinal cord. This remains a revolutionary idea, given the still-present bias in the behavioral sciences for sensory-driven, reflex-based mechanisms of behavior. Preyer's main evidence for motor primacy was that the chick embryo was motile several days before a response (reflex) could be evoked, suggesting that the behavior was spontaneous during this *pre-reflexogenic* period, and probably beyond. He suggested further that the embryonic behavior of other species may be spontaneous, even if they do not have a pre-reflexogenic period.

Experimental support for Preyer's propositions of motor primacy and spontaneity came more than seventy-five years later, mostly from embryologist Viktor Hamburger and his colleagues at Washington University in St. Louis. Using the chick embryo as a model, Hamburger and colleagues confirmed a pre-reflexogenic period for the chick;[8] spontaneous movements appear at three and a half to four days of incubation, about three days before sensory stimuli can evoke the first response on day seven.[9,10] The spontaneous character of behavior at later stages was established using a two-step microsurgical process. First, brain connections with the spinal cord were prevented by transecting the spinal cord at the cervical (neck) region, creating a so-called *spinal embryo*. Second, sensory input to the spinal cord below the level of transection was prevented, creating a *deafferented embryo*.[11] (Connections between brain and sensory nerves were "prevented" instead of "eliminated" because the relatively noninvasive procedures were performed before such structural and functional connections were established.) The motor output of the spinal cord via the ventral roots was not disturbed. The behavior of Hamburger's experimental embryos that lacked both brain and sensory input (deafferented-spinal embryos) was normal in frequency and pattern, indicating that prenatal behavior of the chick is spontaneous and produced by activity within the spinal cord, not the brain.[10]

I joined Viktor Hamburger's research team as an electrophysiologist with the task of discovering the spinal cord activity producing these embryonic movements. The formidable problems of recording nerve impulses from tiny, fragile embryos were eventually solved, and I bagged my neurological quarry. Embryonic movement is driven by massive spinal cord discharges of a type unique to the embryo.[12,13] When a burst occurred in one region of the spinal cord, it recruited adjacent neurons and swept throughout the remainder of the cord—there was no specific site of origin of the discharges within the ventral spinal cord.[14,15] The motor neurons participating in the discharges sent surges of activity via the ventral roots of the cord to muscles that contracted, producing movement.[16] Movements did not occur in the absence of the discharges. The spinal cord (not brain) origin of the discharges was demonstrated by establishing their presence in embryos whose spinal cord was disconnected from the brain using microsurgery.[17]

The spinal cord discharges are a cause—not a consequence—of movement, because they continued unchanged in embryos paralyzed with the notorious drug curare.[18] Amazonian aboriginals first used curare as an arrow-tip poison for hunting, defense, and nailing the occasional missionary. At low levels, curare is a medically useful muscle relaxant. Curare paralyzes (or relaxes) by blocking a muscle's ability to contract in response to squirts of neurotransmitter released by motor neurons that innervate them. Although curare is potentially deadly after birth because it blocks breathing, it does not impede the cardiovascular-based respiration of embryos that gain sustenance through the placenta and umbilical cord. Chick embryos are ideal for such experiments because they respire through the shell, not via pulmonary respiration, and do not have a problematic mammalian maternal host.

Spontaneous spinal cord discharges and associated behavior of the sort found in embryos decline during late development and are not present after birth. The jerks and thrashing of the embryo are totally different from the smoother, goal-directed actions of babyhood and beyond. Indeed, the perseverance of embryonic-type activity into postnatal life would be catastrophic, making life as we know it impossible—we would be flailing about, racked continually by massive, possibly fatal seizures. This tendency of parts of the healthy, immature nervous system to generate massively coupled neuronal activity may be of interest to epilepsy researchers who, using animal models, often produce and then treat seizures that may be more experimental artifact than natural phenomenon. Pro-life advocates seeking evidence for the continuity of prenatal and postnatal life will find little to like here. Embryonic behavior does not provide compelling evidence for sentience and personhood unless these qualities are presumed to reside in the spinal cord and manifest themselves in seizure-like discharges.[19]

~

Embryonic behavior is neglected by developmental psychologists, banished as if it were an errant child not living up to expectations. After all, what's so interesting or psychologically relevant about spontaneous, seizure-driven thrashing that takes place while we are waiting for the good stuff to develop? The brain—that holy of holies of behavioral science—may be minimally involved. Admittedly, embryos are not impressive psychological entities, but their behavior has huge, unappreciated consequences. Consider what happens when movement stops.

Curare-induced paralysis has a catastrophic effect on joint development in chick embryos, causing abnormal "frozen" joints (ankylosis).[20] Movement is necessary to sculpt the male and female components of developing ball-and-socket joints, and malformations result when it is blocked by curare. The well-mated joint surfaces of normally moving embryos are thus not a miracle of genetic programming but the result of them being sculpted against each other during growth. They are molded for you and by you, and pathology of movement can yield pathology of structure.[21] Again, a satisfactory result is produced by a wonderfully effective but sloppy developmental process with a wide tolerance for error.

Movement-related skeletal pathology is common during normal human development.[22] For example, such common congenital anomalies as clubfoot[23] and dislocation of the hip may be caused by restricted movement during the prenatal period, sometimes the result of a malformed uterus.[22] ("Congenital" refers to a trait present at birth, not, as commonly believed, having a genetic basis. In fact, only a few percent of congenital defects have a recognized genetic basis.) Obstetricians will discover other pathology as they develop an eye for such phenomena. Orthopedists fully appreciate the role of postnatal movement, the reason why they are reluctant to immobilize injured joints while they heal. Nonskeletal consequences of embryonic movement include stretching the skin to fit our body (it's too tight without stretching),[24] regulating umbilical cord length (it's too short without stretching),[25] and increasing muscle mass (muscles atrophy without exercise). Movement also influences muscle differentiation (formation of fast-twitch and slow-twitch muscles),[26] size of the motor neuron population (discussed below), and synaptic connections in the nervous system.[27]

~

Behavior is the result of muscle contractions activated by motor neurons. What ensures that we have the necessary number of motor neurons for this critical task? Is our allotment determined genetically or by some other mechanism? This is another of those significant but nontraditional questions overlooked by behavioral scientists. As it turns out, the size of our motor neuron pool comes about in a rather indirect fashion, and behavior plays a central role.

The embryo overproduces spinal cord motor neurons, creating about twice as many as will survive, and then reducing the surfeit through naturally occurring cell death (apoptosis).[28] Death is the embryo's preferred means of neurological population control. The dying of cells during early development is common, although we typically think of development as an additive process, not a subtractive one. Without selective cell death, your body would lack orifices, including the ones you are looking through to read this text, the ones that you breathe through, eat through, listen through, and so on. Without cell death in your interdigital spaces, your hands and feet would be paddles that lack separate fingers and toes—good for swimming, but less suited for grasping or turning the pages of this book.

Motor neuron death (and survival) varies as a function of the amount of muscle tissue to be innervated.[28] For example, unilateral removal (extirpation) of the bud that forms a wing or leg causes the death of most motor neurons on the side of the body that would ordinarily innervate it. (The contralateral side of the embryo with its intact limb bud was the control condition.) Thus, the survival of motor neurons is proportional to the amount of presumed *trophic substance*, a chemical agent necessary for the development and maintenance of cells. The more of a limb bud that is removed, the less trophic substance is available, and the greater the amount of neuron death. The corollary hypothesis that the death of neurons is reduced by adding trophic substance was tested by transplanting an additional (supernumerary) limb bud (and source of trophic substance) near the normal limb bud and counting motor neurons that survived development on that side of the body. (Again, the intact, contralateral limb bud was the control.) As predicted, more motor neurons survived (fewer died) on the side with the supernumerary limb than on the control side with a single limb. Thus, the number of motor neurons that reach maturity is the consequence of a two-step epigenetic process in which an initial oversupply is winnowed through competition at the neuronal level for a limited amount of trophic substance. The half of motor neurons that die are not doomed at their conception. This developmental process ensures that the number of surviving motor neurons will match the amount of limb tissue to be innervated.

~

Especially intriguing is that the massive motor neuron die-off begins when the embryos first move their limbs, suggesting that behavior contributes to motor neuron death. This improbable proposition was tested by paralyzing embryos with curare and counting the number of surviving motor neurons. Remarkably, the 50 percent of motor neurons that die during normal development are saved by paralyzing the embryo.[29] However, the reprieve is short-lived. Death resumes when the paralytic agent wears off and movement returns, decreasing the number of motor neurons to levels typical in normal embryos. Increasing the amount of limb movement by electrical stimulation has the opposite effect of paralysis, increasing the amount of motor neuron death.[30] Given traditional thinking about behavior as the obedient servant of its neuronal master, it's a revelation that behavior development controls the fate of motor neurons. Similar activity-regulated processes may play a role in the survival of brain interneurons that synapse only with other neurons and produce behavior more complex than simple muscle contractions. In ways that we are only starting to understand, the brain produces behavior, and behavior shapes the brain.

Prenatal behavior provides a lesson of contrasts with the postnatal realm—contrasts in form, function and mechanism—that forces a rethinking of the causes and consequences of behavior, and of brain-behavior relations. At every level, prenatal behavior is the quirkiest of the quirky behaviors considered in this book, and provides a fitting conclusion. But there is nothing quirky about the power of developmental analysis. Age-old mysteries about the nature and nurture of human behavior may be clarified, if not solved, by life-span developmental studies starting with the embryo.

Early prenatal development, not birth, childhood, or puberty, is life's most critical stage, and its lessons must be learned from the embryo, not specialists of babyhood and beyond. As observed by a distinguished investigator in another discipline, "Life is infinitely stranger than anything which the mind of man could invent. We would not dare to conceive the things that are really mere commonplaces of existence" (A. Conan Doyle, "A Case of Identity," 1891).

APPENDIX

NOTES

REFERENCES

ACKNOWLEDGMENTS

INDEX

APPENDIX

the behavioral keyboard

The Behavioral Keyboard is a piano-like display that illustrates the relative reaction time of ten common behaviors to verbal command, an estimate of *voluntary* control. (*Involuntary,* reflex responses to appropriate nonverbal stimuli probably have much shorter latencies.) On the keyboard, reaction times range from the sluggish cry (9.8 seconds) on the far left to the rapid eye-blink (0.5 second) on the far right. Reaction time provides a measure of voluntary control of behavior; the shorter the reaction time, the greater the voluntary control. Reaction time also reveals differences in mechanism; acts having very different response latencies involve different neurological processes.

A stopwatch was used to measure the reaction times of 103 participants who were asked to cry, hiccup, sneeze, yawn, laugh, cough, say "ha-ha," inhale, smile, and blink, in randomized order. After saying "Now," the experimenter started the watch, stopping it when the requested act was initiated and recording the reaction time or else recording the maximum

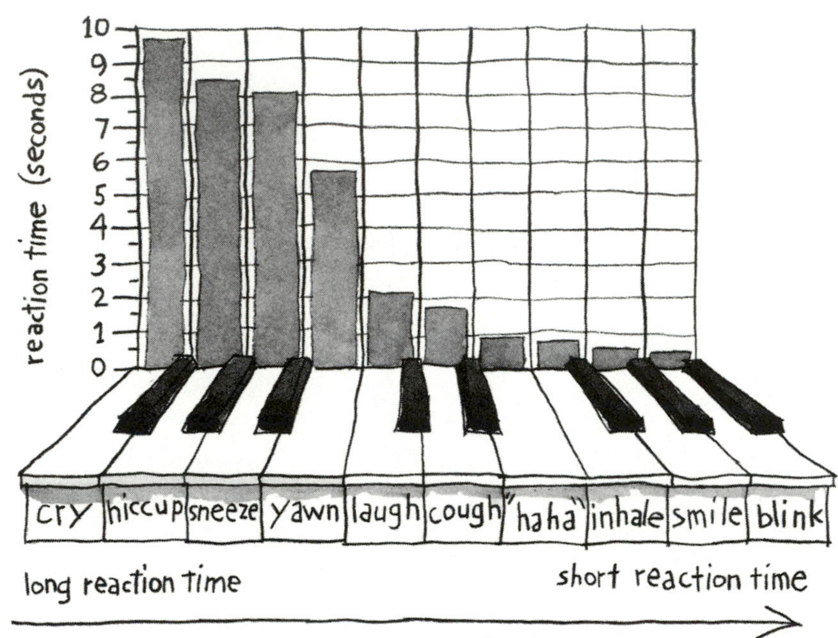

The Behavioral Keyboard summarizes the relative reaction times and associated voluntary control of ten common behaviors. Response latency is inversely related to voluntary control, ranging from the sluggish, hard-to-play vocal cry (left) *to the quick, easy-to-play blink* (right).

of ten seconds if the act could not be performed within the ten-second limit. A manually controlled stopwatch was used to time responses, instead of the more technically sophisticated techniques of video microanalysis or electromyography, because it is less invasive in clinical settings, available to anyone wanting to replicate these results, and satisfactory for establishing relative (not absolute) response latencies.

Reaction times and associated voluntary control varied greatly. Crying was particularly difficult to perform—only 3 percent of participants even attempted it within the allowed ten seconds. Other difficult acts and their completion rates are hiccups (18 percent), sneezes (22 percent), and yawns (58 percent). Success rates are probably overestimated (and reaction times underestimated) because some participants likely faked a behavior to comply with the experimenter's request. In contrast to these challenging acts, all participants were able to say "ha-ha," inhale, smile, and blink within less than one second, evidence of a relatively high level of voluntary control.

Reaction time differentiates social role and mechanism. Both laughs and smiles, for example, are signals of positive affect, but the smile has a much higher degree of voluntary control than the laugh, as reflected in its much shorter reaction time (0.6 versus 2.1 seconds). The voluntary smile offers a more subtle and nuanced social signal than the less frequent, involuntary vocal bludgeon of laughter. The much longer reaction time for laughing (2.1 seconds) than saying "ha-ha" (0.9 second) indicates further that laughing is not a case of speaking "ha-ha," a significant issue in understanding why we laugh (Chapter 2).

The eight airway maneuvers of the Behavioral Keyboard have reaction times ranging from very short to very long. In order of increasing reaction time (and decreasing voluntary control) are inhaling (0.8 second) and saying "ha-ha" (0.9 second), with coughing taking about twice as long (1.7 seconds) as either. Laughing (2.1 seconds) and the real laggards, yawning (5.7 seconds), sneezing (8.1 seconds), hiccupping (8.4 seconds), and crying (9.8 seconds), have much longer reaction times, if they can be performed at all.

Coughs and sneezes, both complementary, explosive maneuvers to clear the airways, have very different reaction times (1.7 seconds for coughing, 8.1 seconds for sneezing) and rates of participation: 95 percent of participants coughed within ten seconds, versus only 22 percent for sneezing. Thus it's easy to cough on command, but difficult, if not impossible, to sneeze on command.

Nicole W. Brocato, Kurt Krosnowski, Clifford Workman, and Megan Hosey assisted in the collection of reaction time data.

notes

1. yawning

1. Provine 1986.
2. Walusinski's edited book (2010) features reviews by leading researchers and is an excellent entrée to contemporary yawn science. His website (www.baillement.com) is a rich source of the often scattered and obscure literature about yawning. Wolter Seuntjens's *On Yawning, or the Hidden Sexuality of the Human Yawn* (2004) is broader than its title suggests and is remarkable in depth and creativity. Provine 2005 provides a brief, accessible review of yawning and its contagion.
3. Provine 1989b.
4. Carskadon 1991, 1992.
5. Provine 1989c. This early attempt to create computer animations using an already antiquated Apple IIe computer and homebrew software became less appealing when it was realized that the yawn detector was broadly tuned and might not respond to abstract visual stimuli (Provine 1989b). Using better technology, Campbell, Carter, Proctor, Eisenberg, and de Waal (2009) pursued a somewhat similar animation tactic in their study of contagious yawning in chimpanzees.
6. Arnott, Singhal, and Goodale 2009.
7. Moore 1942.

8. Platek 2010; Provine 2000; Preston and de Waal 2002.
9. Rizzoletti and Fabbri-Destro 2010; Iacoboni 2009.
10. Ramachandran and Oberman 2006.
11. See Platek 2010, 110. Also see Platek, Mohamad, and Gallup 2005; Schurmann, Hesse, Stephan, Saarela, Zilles, Hari, and Fink 2005; Arnott, Singhal, and Goodale 2009; Nahab, Hattori, Saad, and Hallett 2009; and Nahab 2010.
12. Anderson, Myowa-Yamakoshi, and Matsuzawa 2004. Admission to the club of contagious yawners is granted grudgingly. Chimpanzees, as with humans, yawn contagiously to video images of other yawning chimpanzees (Anderson, Myowa-Yamakoshi, and Matsuzawa 2004), yawn to computer-generated animations of chimpanzee yawns (Campbell, Carter, Proctor, Eisenberg, and de Waal 2009), and yawn more to familiar than unfamiliar individuals (Campbell and de Waal 2011). Evidence for contagion is less clear as we move beyond chimpanzees (Anderson 2010; Deputte 1994). Also see Campbell et al. 2009; Yoon and Tennie 2010; and Campbell and de Waal 2011.
13. Gallup 1970.
14. Paukner and Anderson 2006 report that videotapes of yawning conspecifics (members of the same species) evoke more yawns than control tapes without yawns in stumptail macaques, an Old World monkey, but the effect is confounded by the occasional use of voluntary yawns, with brandished incisors, as a threat gesture in this species. Thus, apparent contagion may be a stress response to a viewed threat gesture, not true contagion as we know it in humans. Contagious yawning has also been reported in gelada baboons, another Old World monkey (Palagi, Leone, Mancini, and Ferrari 2009), but see Anderson 2010.
15. Joly-Mascheroni, Senju, and Shepherd (2008) report that human yawns can trigger contagious yawns in dogs, but Harr, Gilbert, and Phillips (2009) and O'Hara and Reeve (2011) question whether dogs can catch yawns from either humans or other dogs, at least as a higher-order, cognitively based act of empathy. Also see Senju's (2010) response to Harr, Gilbert, and Phillips 2009; Anderson 2010; and Campbell and de Waal's (2010) discussion of methodology.
16. Anderson and Meno 2003; Helt, Eigisti, Snyder, and Fein 2010; Provine 1989a.
17. Senju, Maeda, Kikuchi, Hasegawa, Tojo, and Osanai 2007. Also see Giganti and Esposito Ziello 2009 and Helt, Eigsti, Snyder, and Fein 2010.
18. Senju, Kikuchi, Akchi, Hasegawa, Tojo, and Hiro 2009. The significance of eyes in mediating contagious yawning in autistic individuals complements the work of Provine (1989b), who found that the squinting eyes contribute to contagious yawning of normal adults, and Kiln, Jones, Schultz,

Volkmar, and Cohen (2002), who discovered that autistic individuals preferentially fixate on the mouth region, which conveys little information about yawns.

19. Platek, Critton, Myers, and Gallup 2003.
20. Haker and Rossler 2009.
21. Lehmann 1979.
22. Baenninger 1987.
23. De Vries, Visser, and Prechtl 1982. Spontaneous yawning shows marked developmental trends. Yawning appears around eleven weeks postconceptual age (de Vries, Visser, and Prechtl 1982), remaining relatively unchanged in frequency between twenty and thirty-six weeks, after which it declines until forty weeks (Giganti, Hayes, Cioni, and Salzarulo 2007). The frequency of postnatal yawning is highly variable, probably correlated with sleep patterns (Provine, Hamernik, and Curchack 1987), with the elderly yawning less than young adults, especially during the morning and midafternoon (Zilli, Giganti, and Uga 2008). Contagious yawns are observed only from four to five years onward (Anderson and Meno 2003; Giganti and Esposito Ziello 2009; Helt, Snyder, and Fein 2010).
24. The difficult, if not impossible, sealed-lips nose yawn was first reported in Provine, Hamernik, and Curchack 1987; and Provine, Tate, and Geldmacher 1987; it is also discussed elsewhere (Provine 1996b, 1997b, 2005). The difficulty of the procedure was so obvious that a formal empirical test was deemed unnecessary.
25. Seuntjens 2004.
26. Collins and Equiber 2010.
27. Anderson 2010.
28. Schino and Aureli 1989.
29. Provine, Hamernik, and Curchack 1987.
30. Walshe 1923. Also see Mulley 1982. Meenakshisundaram, Thirumalaikolundusubramanian, Walusinski, Muthusundari, and Sweni 2010 provide an updated review. Typical explanations suggest that associated movements are mediated by spared extrapyramidal pathways, not the damaged pyramidal pathways that mediate voluntary movements. The cause of the movements is unknown, but the reappearance of previously inhibited primitive reflexes is often suggested. The associated movements may be relics of our quadrupedal past when there was a close linkage between locomotion and respiratory acts such as yawning. Chapter 2, "Laughing," and Provine 1996a and 2000 consider the role of bipedality in the unlinking of breathing and locomotion and its role in the evolution of human laughter and speech.

31. Walusinski (2010) reviews yawning in locked-in syndrome and related disorders, including empirical reports by Bauer, Krasnianski, and colleagues. A significant theme in this literature is the relative sparing of involuntary, emotional acts such as yawning and laughing.
32. Heusner 1946 reviews early work of Gamper, Catel, and Krauspe.
33. Provine and Hamernik 1986.
34. This news item was brought to my attention by Gordon Gallup.
35. Provine, Tate, and Geldmacher 1987.
36. Provine 1986, 2005; Provine and Hamernik 1986; Provine, Hamernik, and Curchack 1987; Baenninger 1997; Seuntjens 2004.
37. Baenninger 1997. Gallup and colleagues (e.g., 2007, 2010) offer a fresh entrant in the yawn function sweepstakes—brain cooling and thermoregulation. Their interesting position is based on often indirect experiments and scholarly reviews of brain temperature and yawning in health and disease. For a more direct tack, see Shoup-Knox, Gallup, Gallup, and McNay 2010. Guggisberg, Mathis, Schnider, and Hess 2010 and Guggisberg, Mathis, and Hess 2010 challenge the mostly circumstantial evidence supporting physiological hypotheses of yawn function, including the popular view that yawning enhances vigilance, emphasizing instead its social role. Yawning, they note, does not produce autonomic or EEG indices of arousal and increased vigilance.
38. Eibl-Eibesfeldt 1975, 163.
39. Hatfield, Cacioppo, and Rapson 1994.

2. laughing

1. Rankin and Philip 1963.
2. Provine 1996a; 2000, 130–131.
3. Hemplemann 2007, 52, reviews the laughter epidemic, noting that "Provine 2000: 130–131 . . . presents an accurate summary of Rankin and Philip 1963," but he faults the interpretation, especially the failure to observe that the epidemic was more appropriately a "contagiousness of hysteria" than of laughter. However, if Hemplemann had continued reading Provine 2000, 131–133, he would have discovered that the very next paragraph considered "mass hysteria" and the "broader implications of the phenomenon." Hemplemann usefully observes that the epidemic did not feature nonstop laughing (of course not), that laughter was not caused by humor (of course not), that mass hysteria was associated with stress, times of change, and

cultural factors (yes, a common correlate of mass hysteria), and that crying, another expression of strong emotion was also observed (see Chapter 3). Provine 2000 and this book are full of examples of such human herd behavior, both normal and pathological.

4. Provine 2000, 133–137.
5. Provine 2000, 193–194.
6. Provine 2000, 137–143. Emperor Nero (AD 37–69), an avid actor and student of the Greek theater, even brought his own personal cheering section in the form of five thousand Roman soldiers to applaud his performances, for which the wise judges always awarded first prize. The use of such claques (from the French *claquer,* "to clap") continues in opera, with Hector Berlioz, composer and sometimes music critic, referring to claques as "Romans."
7. Provine 2000, 143–147.
8. Provine 1992; Smoski and Bachorowski 2003.
9. Chimpanzees also produce laugh-elicited laughter, but the vocalization differs in acoustic form and occurrence from their spontaneous laughter (Davila-Ross, Allcock, Thomas, and Bard 2011). Great apes laugh during social play and tickling, where it prolongs play bouts (Matsuasaka 2004; Vettin and Todt 2005). Laughter (Rothbart 1973) and humor may play a similar role in humans (Weisfeld 1993).
10. Bachorowski and Owren (2001) found that laughter is rated as more pleasurable if varied in structure and less repetitive.
11. Provine and Yong 1991; Provine 1992, 1996a, 1996b, 1997a, 1997b; Provine 2000, 149–151.
12. Provine 2000. Also, Sauter, Eisner, Ekman, and Scott (2010) report that "laughter was cross-culturally recognized as signaling joy" (2411).
13. Provine 2000, 49–53.
14. Gervais and Wilson (2005) differentiate between spontaneous, emotion-driven Duchenne laughter and voluntarily controlled non-Duchenne laughter. Human laughter is a mix of both varieties, with other apes probably restricted to the honest, Duchenne variety.
15. Provine and Yong 1991; Provine 2000, 56–57.
16. Provine and Yong 1991; Provine 2000, 55–64.
17. Provine 2000, 57–59. Bachorowski, Smoski, and Owren (2001) challenge the report that "ha," ho," and "he" are common laugh variants. Their point is irrelevant to Provine's central thesis that laugh notes (syllables, calls, etc.), however described, have distinct acoustic properties, and that there are vocal constraints on producing such difficult laugh variants as "ha-ho-ha-ho" versus the simple "ha-ha-ha-ha."

18. "Stereotyped" was carefully selected as a term to describe the strong central tendency around which laughter varies, not "fixed action pattern" or other suggestion of rigidity or invariance. Provine (2000, 63) states, "Our nervous system and vocal track enforce this stereotypy—we can't laugh in arbitrary ways even if we try," documenting the point with some challenging vocal gymnastics (e.g., "ha-he-ha-he") and continuing, "Although laughter is stereotyped, it's not totally inflexible. People laugh in different ways at different times, expressing social, grammatical, and emotional nuance." Bachorowski and Owren (2001) muddy the waters by titling a paper "Not All Laughs Are Alike," when no one takes this position, certainly not Provine. Their treatment of behavioral stereotypy is reminiscent of early attacks on ethology concerning "how fixed is a fixed action pattern." Their paper provides a sophisticated and useful acoustic analysis of laughter variants, noting, for example, that voiced, song-like laughs were more favorably received than variants that were unvoiced grunts, pants, and snort-like sounds. Grammer and Eibl-Eibesfeldt (1990) also distinguish between voiced (vocalized) and unvoiced laughter. The challenge to stereotypy continues in Bachorowski, Smoski, and Owren (2001), who note, "We found laughter to be a repertoire of highly variable vocalizations that includes qualitatively distinct voiced song-like, unvoiced grunt-like, and unvoiced snort-like versions" (1594). Also, "these results stand in contrast to claims that laughter is a stereotyped vocal signal and highlight the difficulty of trying to characterize laughter as being a single acoustic form" (1596). Although they make the obvious point that "all laughter is not the same," we must revisit a central issue of Provine (2000) and this chapter: *without some underlying invariance, a vocalization could not be identified as laughter.*
19. Van Hoof (1972) hypothesizes that human smiling is homologous to the ape "silent bared-teeth display" (grin) produced under threat, and that human laughter is homologous to the "relaxed open-mouth display" (play face), accompanied by play panting in some species. The focus on the play face differs from that of Provine (2000) and others (e.g., Davila Ross, Owren, and Zimmermann 2009) who attend to the evolution of the acoustic component of laughter. A problem with the facial focus is that it may not accompany all ape laughs, and possible antecedent laugh forms such as those found in rats by Panksepp (2007) may totally lack the facial component.
20. Darwin 1872; Fossey 1972; Goodall 1986.
21. Provine 2000, 77–79.
22. Provine and Yong 1991.

23. Provine 1996a; 2000, 81–92. Davila Ross, Owren, and Zimmermann (2009) added gorillas, orangutans, and bonobos to the research mix of chimpanzees and humans, providing valuable contrasts between the tickle-induced laughter of a variety of great apes and humans. Davila Ross confirmed Provine's (1996a, 2000) observations that humans produce exclusively "egressive" laughter (during exhalation), while chimpanzees produce primarily alternating "egressive-ingressive" laughter (during exhalation-inhalation), and that humans produce voiced, vowel-like laughter in contrast to the noisy, unvoiced, chimpanzee variant. Other great apes (orangutans, gorillas, bonobos) were found to produce both alternating egressive-ingressive and egressive laughter of the noisy, unvoiced ape variety, but the bonobo had occasional hints of voicing. Gorillas and bonobos, unlike chimpanzees, were reported to sustain egressive airflow longer than their regular breath cycle, suggesting that their vocal competence is less constrained than suggested by Provine (1996a, 2000, this chapter). The suggestion by Davila Ross that "evolutionary changes occurred along existing dimensions of variation, rather than being de novo inventions" (1107) is not inconsistent with Provine (2000, this chapter), who stresses the transformative role of bipedality and associated neurophysiological and neuroanatomical changes (e.g., Bramble and Currier 1983; MacLarnon and Hewitt 1999; McFarland 2001). Bipedality frees the thorax of its mechanical support function and permits the emergence of vocal variants on which selection can operate, accelerating the velocity of vocal evolution. The fact remains that, compared to humans, apes laugh in a different way and are seriously constrained in vocal production, and their laughter probably originated in the ritualized panting of tickle and rough-and-tumble play. Also see van Hoof and Preuschoft 2003; Vettin and Todt 2005; Provine and Bard 1994, 1995; and Provine 1999.
24. Provine 2000, 84–92.
25. Bramble and Currier 1983.
26. Sroufe and Waters 1976; Sroufe and Wunch 1972. As with chimpanzees (Plooij 1979), human babies laugh when tickled/touched by their mothers, a vocalization that prompts more maternal contact and baby laughter until the baby is overstimulated and fusses, causing the mother to stop.
27. Panksepp 2007; Panksepp and Burgdorf 2003.
28. Provine 2000, 92–97; Roger Fouts, personal communication.
29. Provine and Fischer 1989.
30. Provine 1993; Vettin and Todt 2004.
31. Although this chapter is primarily about laughter, not humor, the study of both benefits from Rod Martin's scholarly text *The Psychology of Humor*

(2007) and the just-published *Inside Jokes* (Hurley, Dennett, and Adams 2011).

32. Provine 2000, 27–32; Grammer 1990; Grammer and Eibl-Eibesfeldt 1990.
33. Susan Prekel in Nicholson 2010, 38.
34. Provine 2000, 32–35.
35. Provine 1993; 2000, 36–39.
36. Bramble and Currier 1983; Winkworth, Davis, and Adams 1995; MacLarnon and Hewitt 1999; McFarland 2001.
37. Provine and Emmorey 2006. The finding that vocal laughter punctuates the conversation of deaf signers indicates that the punctuation process involves a higher-order cognitive and linguistic process because, unlike vocal speech and vocal laughter, the two motor acts do not compete for the same vocal output channel. Makagon, Funayama, and Owren (2008) show that laughter of congenitally deaf college students is essentially similar to that of normally hearing individuals, but had lower amplitude and longer duration.
38. Provine in preparation. The constant contact via cell phones with a preselected set of friends in a global arena provides an emotionally rich and satisfying exchange, but it comes at the expense of contact with people physically present. The phenomenal success of modern communication technology, from cell phones to the social medium of Facebook, is due more to social engagement than utilitarian messaging. The task of small talk and gossip is bonding, and a relationship benefits from the mere act of talking.
39. Provine, Spencer, and Mandell 2005.
40. Provine 2000, 189–207, reviews medicinal laughter and humor. For recent research on medicinal effects, see Dunbar, Baron, Frangou, Pearce, van Leeuwen, Stow, Partridge, MacDonald, Barra, and van Vugt (in press), about analgesic effects of social laughter. Miller and colleagues (e.g., Miller and Fry 2009) report a direct effect of laughter on dilating (expanding) the endothelium (inner lining of blood vessels), mediated by the release of nitric oxide, the kind of heart-protective response associated with aerobic exercise and such medications as statins and ACE inhibitors. A challenge for research about health benefits is in isolating the often confounded effects of laughter, humor, positive mood, and the social context essential for laughter.
41. Friedman, Tucker, Tomlinson-Keasey, Schwartz, Wingard, and Criqui 1993; Gruber, Johnson, Oveis, and Keltner 2008; Gruber, Mauss, and Tamir 2011.

3. vocal crying

1. "Baby cry" is rated the third most horrible sound in the Bad Vibes Web-based survey of more than a million respondents, and one of the rare sounds that males rated worse than females.
2. Ostwald 1972.
3. Vuorenkowski, Wasz-Hockert, Koivisto, and Lind 1969.
4. Mead and Newton 1967.
5. Soltis 2004. The lead article, commentaries, and author's rebuttal provide a rich and varied sample of contemporary thought about crying as a signal.
6. Bowlby 1982; Bell and Ainsworth 1972; Christensson, Cabrera, Christensson, Uvnas-Mosberg, and Winberg 1995.
7. Barr, Paterson, MacMartin, Lehtonen, and Young 2005.
8. Barr, Konner, Bakeman, and Adamson 1991.
9. Spock 1968.
10. Barr and Trent 2006.
11. Wolff 1969, 1987; Brazelton 1962; Barr, Hopkins, and Green 2000.
12. Barr, Chen, Hopkins, and Westra 1996.
13. Hunziker and Barr 1986.
14. Frey 1985.
15. Gekowski, Rovee-Collier, and Carulli-Rabinowitz 1983.
16. Gessell and Ilg 1943.
17. Wessel, Cobb, Jackson, Harris, and Detwiler 1954.
18. Buhler and Hetzer 1928.
19. Simner 1971.
20. Geangu, Benga, Stahl, and Triano (2010) provide a brief, up-to-date review of the development of contagious crying and empathy.
21. Martin and Clark 1982.
22. Dondi, Simion, and Caltran 1999. See also Sagi and Hoffman 1976 and Field, Diego, Hernandez-Reif, and Fernandez 2007.
23. Spinrad and Stifter 2006.
24. Zahn-Waxler, Friedman, and Cummings 1983; Hoffman 2000; Decety and Jackson 2004.
25. Freeman 1964; Eibl-Eibesfeldt 1973.
26. Sroufe and Waters 1976.
27. Gessell and Amatruda 1945.
28. Provine and Yong 1991.
29. Provine 2000.
30. Provine 1996a.

31. Provine 1992.
32. Provine 1986.
33. Provine 2005.
34. Provine 2000. For a broad perspective of pathological laughing and crying, see Provine 2000, Chapter 8, "Abnormal and Inappropriate Laughter." Parvizi and colleagues (e.g., Parvizi, Anderson, Martin, Damasio, and Damasio 2001) are the most active recent researchers of the neurology of pathological laughing and crying. They emphasize the role of the cerebellum and associated structures in adjusting the execution of laughing and crying to fit cognitive and situational context.
35. Frances McGill, in Lieberman and Benson 1977.

4. emotional tearing

1. Frey 1985.
2. Sullivan, Stern, Tsubota, Dart, Sullivan, and Bloomberg 2002.
3. Lutz 1999.
4. Van Haeringen 2001.
5. Provine 2009.
6. Vingerhoets and Cornelius 2001.
7. Masson and McCarthy 1995.
8. Darwin 1872; Hopkins 2000; Apt and Cullen 1964; Isenberg, Apt, McCarthy, Copper, Lim, and Signore 1998.
9. Rottenberg, Bylsma, and Vingerhoets 2008; Kraemer and Hastrup 1988.
10. Oleshansky and Labbate 1996.
11. Gelstein, Yeshurum, Rosenkrantz, Shushan, Frumin, Roth, and Sobel 2011.
12. Provine 2011.
13. Lambiase, Rama, Boini, Caprioglio, and Aloe 1998.
14. Croassin, Lambiase, Costa, DeGregorio, Sgrulletta, Sacchetti, Aloe, and Bonini 2005.
15. Two lines of work suggest that NGF may have an effect on emotional experience as well as emotional expression. Given that tears contain NGF, and that NGF has an antidepressant effect (Altar 1999; Angelucci, Mathe, and Aloe 2004; Duman and Monteggia 2006), at least in rats, it follows that human emotional tears, if bearing NGF, may modulate mood. In humans, conditions are ideal for the introduction of tear-related NGF directly into the brain. During weeping, tears produced by the lacrimal glands drain through the nasocranial duct into the nasal cavity (the reason why teary individuals have runny noses). The NGF in these tears can

bypass the blood-brain barrier and gain rapid extracellular and intracellular access to the brain via the olfactory and trigeminal nerves (Benedict, Frey, Schioth, Schultes, Born, and Hallschmid 2011). The present chapter is the first proposal that NGF may have an antidepressant effect via emotional tears. The possibility of antidepressant emotional tears prompts a rethinking of the consequences of weeping and its modification by drug use or pathology.

16. Levi-Montalcini 1988.
17. Ekman, Friesen, and Ellsworth 1972.
18. Pentland 2008.
19. Research on facial expressions also suffers from this affliction, treating smiling, for example, as a static posture instead of a neuromuscular event distributed in time. The description of the microstructure of facial postures in terms of the electromyography of specific muscles acknowledges the complexity of facial behavior but provides only a high resolution neuromuscular snapshot of a dynamic behavior.
20. It's informative to contrast painting, a time-free medium of space, with music, a space-free medium of time. A painting is a pattern of varying shapes, brightnesses, and colors distributed in space. You can take a "snapshot" of the *Mona Lisa* and the image of this instantaneous sample is completely recognizable up to the point where the sample is so short that it encounters issues with the quantal property of light, becoming an increasingly fuzzy snowstorm of photons. In contrast, you can't take a comparable instantaneous acoustic sample (snapshot) of "Twinkle Twinkle Little Star" because melodic structure depends upon a sequence of pitches of varying loudness that unfolds in time. An instantaneous sample of a tone would itself be inaudible; lacking frequency, it would be only a pressure, such as 15.7 pounds per square inch. Music, because of its time structure, demands more of its audience than visual art, a reason why modern art may be accepted more readily than modern music. You can briskly walk through a gallery or flip through a book of images, pausing to examine what you like. You can't fast-forward through a concert. A live musical performance is a commitment. Opera is multimodal, industrial-strength Italian melodrama set to song. Dance is a dynamic visual art of pure movement—sculpture in motion, usually accompanied by music. Radio drama is theater of the mind, a purely auditory art form of the fourth dimension. Music, an even more abstract, austere, auditory art, can evoke emotion without direct enactment of human voice or movement.

5. whites of the eyes

1. Blank, Provine, and Enoch 1975.
2. Provine and Enoch 1975.
3. Provine, Cabrera, Brocato, and Krosnowski 2011.
4. Leibowitz 2000.
5. Murphy, Lau, Sim, and Woods 2007.
6. Papavramidou, Fee, and Christopoulou 2007.
7. Mueller and McStay 2008.
8. Donshik 1988.
9. Owen, Newsom, Rudnicka, Ellis, and Woodward 2005.
10. Paton 1961.
11. McLane and Carroll 1986.
12. Sloan 1962.
13. Roche and Kobos 2004.
14. Broekhuyse 1975.
15. Watson and Young 2004.
16. Gangestad and Thornhill 1999.
17. Johnston 2006.
18. Langlois, Kalakanis, Rubenstein, Larson, Hallamm, and Smoot 2000.
19. Little, Apicella, and Marlowe 2007.
20. Rhodes 2006.
21. Sugiyama 2004, 2005.
22. Thornhill and Gangsestad 1999.
23. Tomasello, Hare, Lehmann, and Call 2007.
24. Etkoff 1999.
25. Symonds 1979, 1995.
26. Habitual users of eyedrops, a group that includes stoners, may experience prolonged eye redness because of dilated blood vessels. Red eye may also be a sign of a more serious underlying condition that is not treated by temporarily reducing blood flow to the eye's surface.
27. Hess 1975.
28. Yanoff 1969.
29. Norn 1985.
30. Kobayashi and Kohshima 2001.
31. Perrett and Mistlin 1991.
32. Emery 2000.
33. Haxby, Hoffman, and Gobbini 2000.
34. Burnham 2003.

35. Haley and Fessler 2005.
36. Bateson, Nettle, and Roberts 2006.
37. Burnham and Hare 2007.

6. coughing

1. Clerf 1947.
2. Irwin, Curley, and French 1990.
3. Criley, Blaufuss, and Kissel 1976; Davis 1983; Criley, Niemann, Rosborough, and Hausnecht 1986.
4. American Heart Association 2011.
5. Girsky and Criley 2006.
6. Petelenz et al. 1998.
7. Lockey, Poots, and Williams 1975.
8. Bering 1955; Du Boulay, O'Connell, Currie, Bostic, and Verity 1972.
9. Kerr and Eich 1961.
10. McFarland 2001; Winkworth, Davis, Adams, and Ellis 1995.
11. Bailey 2008.
12. Heath 1989.
13. Pennebaker 1980.

7. sneezing

1. Sadanand, Kelly, Varughese, and Forney 2005.
2. Vogel 1979.
3. Murray and Bierer 1951.
4. Gallia and Roscoe 1981.
5. Hersch 2000.
6. Batsel and Lines 1975; Stromberg 1975; Nonaka, Unno, Ohta, and Mori 1990; Seijo-Martinez, Varela-Freijanes, Grande, and Vazquez 2006.
7. Everett 1964; Whitman and Packer 1993.
8. Bhutta and Maxwell 2008.

8. hiccupping

1. Fesmire 1988.
2. Odeh, Bassan, and Oliven 1990.
3. Peleg and Peleg 2000.

4. Mayo 1932.
5. Anthoney, Anthoney, and Anthoney 1974.
6. Anthoney, Anthoney, and Anthoney 1978.
7. Anthoney 2010. In a telephone conversation and several emails in February and March 2010, Terence Anthoney generously shared his vast and unique clinical experience and knowledge about hiccupping.
8. de Vreis, Visser, and Prechtl 1985.
9. Schmidt, Cseh, Hara, and Kubli 1984.
10. de Vries, Visser, and Prechtl 1982.
11. de Vries 2011; personal communication.
12. Pillai and James 1990.
13. Souadjian and Cain 1968.
14. Fisher 1967. Intractable hiccups are far more common in males than females, usually occurring in individuals older than forty-five. This suggests the disinhibition of hiccups by decreasing androgen levels in aging males, but it leaves unanswered the relative sparing of females, who always had lower levels of androgens.
15. Lewis 1985.
16. Launois, Bizec, Whitelaw, Cabane, and Derenne 1993.
17. Friedman 1996.
18. Sarkies 1921.
19. Newsom-Davis 1970.
20. Plum 1970.
21. Vincent 2010.
22. Fuller 1990.
23. Straus 2003. The evolutionary scenario of Christian Straus and colleagues is that hiccups are a remnant of the gill breathing of tadpoles, a kind of gulping of air and water. Supporting their position is evidence that both hiccups and amphibian gulps are inhibited by elevated CO_2 and the drug baclofen. I'll join the evolutionary queue and provide my own proposal—that the primal gulp also is an ingestion reflex in which the sudden opening of the mouth creates a vacuum that sucks a morsel into the gaping craw. I recall this response from a tedious feeding routine in an embryology lab in which I would dangle tiny worms before the mouths of larval salamanders to entice them to snap up my offering. Their gulping resembled a hiccup but was not repetitive and may not involve the diaphragm. This aquatic response would not be adaptive in the lighter medium of air, in which the only thing sucked in during our hiccups is air. And while contemplating the developmental and phylogenetic beginnings of hiccup-

ping, it's worth considering some ends: for many people, gasping is the last form of respiration before dying.

24. Golomb 1990. In her argument against any hiccup function, Golomb notes that the universality and developmental profile of hiccupping is no more compelling than that for bed-wetting and drooling.

9. vomiting and nausea

1. Rozin and Fallon 1987.
2. Rozin, Haidt, and McCauley 2000.
3. Lang and Sarna 1989.
4. Scott and Verhagen 2000.
5. Dethier 1963.
6. Hoebel, Rada, Mark, and Pothos 1999.
7. Rozin and Kalat 1971.
8. Seligman 1971.
9. Rozin, Millman, and Nemeroff 1986.
10. Stevenson 1999.
11. Klesius 2009.
12. Treisman 1977.
13. Small and Borus 1983.
14. Selden 1989.
15. Moffatt 1982.
16. Levine 1977.
17. Small and Nicholi 1982.
18. Hefez 1985.
19. Rockney and Lemke 1992.
20. Rosenberg 2009.
21. Nemery, Fischer, Boogaerts, Lison, and Willems 2002.
22. Aldous, Ellam, Murray, and Pike 1994.
23. Andritzky 1989.

10. tickling

1. The annual Ig Nobel Prizes are not a dishonor, but typically a good-humored recognition of the unusual, imaginative, and funny in science, medicine, and technology.
2. Provine 2000, 99–127. This is the most detailed account of Provine's questionnaire study and all aspects of tickling.

3. Classical reflexes such as the patellar tendon (knee jerk) reflex can be evoked by the subject or anyone else, and they are short in duration, simple in structure, proportionate to the amplitude of the stimulus, and relatively independent of social context. Tickling, in contrast, occurs in social context, can't be self-stimulated, and is long in duration; its huddling/fending-away/escaping/laughing is complex in structure and varies with context, and its amplitude is not as directly proportionate to the amplitude of the stimulus.
4. Panksepp and Burgdorf 1999.
5. Poole 1998; personal communication.
6. Plooij 1979.
7. Harris 1999.
8. Provine 2004.
9. Ramachandran and Blakeslee 1998.
10. Grandin 1995.
11. Aristotle 1922.
12. Blakemore, Wolpert, and Firth 1998. For an earlier approach to this problem using a mechanical device, see Weiskranz, Elliot, and Darlington 1971.
13. Provine and Westerman 1979.

11. itching and scratching

1. Gewande 2008.
2. Robinson, Szewczyk, Haddy, Jones, and Harvey 1984.
3. Niemeier, Kupfer, and Gieler 2000.
4. Gieler, Niemeier, Brosig, and Kupfer 2002.
5. Papoiu, Wang, Coghill, Chan, and Yosipovitch 2011.
6. Kalat 2009.
7. Johanek et al. 2007.
8. Davidson and Giesler 2002.
9. Andrew and Craig 2001.
10. Handwerker, Forster, and Kirchoff 1991.
11. Schmelz, Schmidt, Bickel, Handwerker, and Torebjork 1997.
12. Davidson, Zhang, Khasabov, Simone, and Giesler 2009.
13. Sun and Chen 2007.
14. Sun, Zhao, Meng, Yin, Liu, and Chen 2009.
15. Sherrington 1906.
16. Stein 1983.

12. farting and belching

1. Nohain and Caradec 1967.
2. Given Mozart's passion for the scatological, he may have found Le Pétomane irresistible. He did, after all, compose "Leck mich im Arsch" (K.231), literally "Lick my ass," a canon in B-flat, bowdlerized after his death into the less colorful "Let us be glad!"
3. Hippocrates 1981.
4. Levitt 1984.
5. Firth 2004, 473.
6. Gutmann 1926; Raum 1939.
7. Montaigne 1958.
8. Wynne-Jones 1975.
9. Pliny 1951.
10. Bigard, Gaucher, and Lasalle 1979.
11. Wilson, Batty, and Dill 2003.
12. See guinessworldrecords.com/records/human_body/extreme_bodies/loudest_burp (accessed July 14, 2011).

13. prenatal behavior

1. Oppenheim 1984; Provine 1988.
2. For books reviewing behavioral embryology, see Gottlieb 1973; Smotherman and Robinson 1988; and Piontelli 2010.
3. Schoenwolf, Bleyl, Brauer, and Francis-West (2009) provide an accessible, illustrated introduction to prenatal human development.
4. Adult humans are capable only of the modest feat of wound healing, having lost the power of regeneration present earlier in life. Adults may be too large and too many developmental branch points removed from the embryonic state to permit the dynamic of redevelopment that is the basis of regeneration. The remarkable ability of salamanders and some other animals to regenerate an amputated structure such as a forelimb involves the regression of cells at the stump of the limb, formation of a blastema, and redevelopment of the lost structure.
5. de Vries 1982; Luchinger, Hadders-Alegra, Van Kan, and de Vries 2008.
6. Piontelli 2010.
7. Gottlieb 1973.
8. The chicken embryo is one of the classical organisms for experimental embryology research. Fertile eggs are easy to obtain and incubate in the

lab; the embryos are fairly large, easy to visualize, and hardy enough to survive surgical manipulations; and the embryos have a nervous system that approximates our own in organization. If chicken behavior seems unimpressive, consider that it is a biped with sophisticated sensory and motor systems. Yes, chickens lack opposable thumbs, can't talk, and are deficient in tool use, but can you fly?

9. Hamburger 1963.
10. Hamburger, Wenger, and Oppenheim 1966.
11. Potential brain input to the spinal cord was eliminated by removing a segment of the neural tube, the structure that develops into the spinal cord. All potential descending and ascending connection with the brain were eliminated for spinal cord regions caudal (tailward) to the gap. Deafferentation (removal of sensory input) was accomplished by removing the dorsal half of the neural tube caudal to the spinal gap, as well as the neural crest, whose cells differentiate into the sensory ganglia and dorsal roots that innervate the spinal cord. These operations were conducted shortly after the closing of the neural tube and before the developing spinal cord received either brain or sensory input. Unlike procedures performed at later stages, they do not deprive the embryo of already established nerve connections.
12. Provine, Sharma, Sandel, and Hamburger 1970. The spinal burst discharges discovered by Provine and reported in a series of papers were subsequently examined by Landmesser and colleagues (e.g., Landmesser and O'Donovan 1984).
13. Provine 1972.
14. Provine 1971.
15. In a series of studies, Ann Bekoff and colleagues (e.g., Bekoff, Stein, and Hamburger 1975) used electromyography to describe the motor output of chick embryos, emphasizing the early appearance of structured output of motor neurons to muscles, such as the reciprocal inhibition of antagonistic muscles acting on a joint. Organized output was found to occur early and to increase in precision during development. Unfortunately, they made no parallel effort to relate the relative precision of muscle activity with the unique, massive, presumably imprecise spinal discharges driving it. Bekoff's research complemented that of Provine (e.g., 1972), but she stressed developmental continuity, while he focused on features unique to the embryo. The precocial coordination of motor output to muscles is not found when the scope of inquiry is enlarged from joints to limbs (Provine 1980). For example, the wings and legs of young chick embryos move in an unorganized way, with a wing moving no more often with the contralat-

eral wing than with the ipsilateral leg. With development, the wings increasingly move in phase, as in flapping, and the legs move increasingly 180 degrees out of phase, as in walking.

16. Ripley and Provine 1972.
17. Provine and Rogers 1977.
18. Provine 1973.
19. The brain and the spinal cord are traditionally considered to be the two parts of the vertebrate central nervous system, a categorization that distorts thinking about their origin, structure, and function. From the developmental perspective, the brain is simply one end of the neural tube, the embryonic structure producing both the brain and the spinal cord. During development, the proliferation of neurons in the spinal cord fills the lumen of the tube, leaving few traces of its genesis. In the mature brain, the lumen of the neural tube survives as the fluid-filled cavities of the ventricles. The greater mass and complexity of the brain, relative to the spinal cord, is the result, in part, of prolonged proliferation of neurons.
20. Drachman and Sokoloff 1966. See Jago 1970 for an incidental human analogue.
21. Gomez, David, Peet, Vico, Chenu, Malaval, and Skerry 2007; Osborne, Lamb, Lewthwaite, Dowthwaite, and Pitillides 2002.
22. Moessinger 1988.
23. Drachman and Coulumbre 1962.
24. Smith 1981.
25. Miller, Higginbottom, and Smith 1981; Moessinger, Blanc, Marone, and Polsen 1982.
26. Salmons and Streter 1976.
27. Firth, Wang, and Feller 2005; Turrigiano 2004; Wong 1999.
28. Hamburger and Oppenheim (1982) review target-dependent programmed cell death in the developing nervous system. Oppenheim, Caldero, Esquerda and Gould (2001) provide a complementary review of target-independent programmed cell death.
29. Pittman and Oppenheim 1978.
30. Oppenheim and Nunez 1982.

references

Aldous, J. C., Ellam, G. A., Murray, V., and Pike, G. 1994. An outbreak of illness among schoolchildren in London: Toxic poisoning not mass hysteria. *Journal of Epidemiology and Community Health, 48*, 41–45.

Altar, C. A. 1999. Neurotropins and depression. *Trends in Pharmacology, 20*, 59–61.

American Heart Association. 2011. Cough CPR. Last modified October 6, 2011. www.heart.org/HEARTORG/Conditions/More/CardiacArrest/Cough-CPR_UCM_432380_Article.jsp#.T0bgi3l7BLU.

Anderson, J. R. 2010. Non-human primates: A comparative developmental perspective on yawning. In O. Walusinski (ed.), *The mystery of yawning in physiology and disease*, 63–76. Basel: Karger.

Anderson, J. R., and Meno, P. 2003. Psychological influences on yawning in children. *Current Psychology Letters, 11*. http://cpl.revues.org/index390.html.

Anderson, J. R., Myowa-Yamakoshi, M., and Matsuzawa, T. 2004. Contagious yawning in chimpanzees. *Proceedings of the Royal Society B*, 271, S468–S470.

Andrew, D., and Craig, A. D. 2001. Spinothalamic lamina I neurons selectively sensitive to histamine: A central neural pathway for itch. *Nature Neuroscience, 4*, 72–77.

Andritzky, W. 1989. Sociopsychotherapeutic functions of ayahuasca healing in Amazonia. *Journal of Psychoactive Drugs, 21*, 77–89.

Angelucci, F., Mathe, A. A., and Aloe, L. 2004. Neurotrophic factors and CNS disorders: Finding rodent models of depression and schizophrenia. *Progress in Brain Research, 146,* 151–165.

Anthoney, T. R., Anthoney, S. L., and Anthoney, D. J. 1974. The human hiccup: Time relationships and ethological significance. In L. E. Scheving, F. Halberg, and J. E. Pauly (eds.), *Chronobiology,* 531–534. Tokyo: Igaku Shoin.

——. 1978. On temporal structure of human hiccups: Ethology and chronobiology. *International Journal of Chronobiology, 5,* 477–492.

Apt, L., and Cullen, B. F. 1964. Newborns do secrete tears. *Journal of the American Medical Association, 189,* 951–953.

Aristotle. 1922. *An Aristotelian theory of comedy with an adaptation of the Poetics and a translation of the Tractatus Coislinianus.* New York: Harcourt, Brace.

Arnott, S. R., Singhal, A., and Goodale, A. 2009. An investigation of auditory contagious yawning. *Cognitive, Affective, and Behavioral Neuroscience, 9,* 335–342.

Bachorowski, J.-A., and Owren, M. J. 2001. Not all laughs are alike: Voiced but not unvoiced laughter readily elicits positive affect. *Psychological Science, 12,* 252–257.

Bachorowski, J.-A., Smoski, M. J., and Owren, M. J. 2001. The acoustic features of human laughter. *Journal of the Acoustic Society of America, 110,* 1581–1597.

Baenninger, R. 1987. Some comparative aspects of yawning in *Betta splendens, Homo sapiens, Panthera leo,* and *Papio sphinx. Journal of Comparative Psychology, 101,* 349–354.

——. 1997. On yawning and its functions. *Psychonomic Bulletin and Review, 4,* 198–207.

Bailey, J. V. 2008. Could patients' coughing have communicative significance? *Communication and Medicine, 5,* 105–115.

Barr, R. G., Chen, S., Hopkins, B., and Westra, T. 1996. Crying patterns in preterm infants. *Developmental Medicine and Child Neurology, 38,* 345–355.

Barr, R. G., Hopkins, B., and Green, J. A. (eds.). 2000. *Crying as a sign, a symptom, and a signal: Clinical, emotional and developmental aspects of infant and toddler crying.* London: MacKeith Press.

Barr, R. G., Konner, M., Bakeman, R., and Adamson, L. 1991. Crying in !Kung San infants: A test of the cultural specificity hypothesis. *Developmental Medicine and Child Neurology, 33,* 601–610.

Barr, R. G., Paterson, J. A., MacMartin, L. M., Lehtonen, L., and Young, S. N. 2005. Prolonged and unsoothable crying bouts in infants with and without colic. *Journal of Developmental and Behavioral Pediatrics, 26,* 14–23.

Barr, R. G., and Trent, R. B. 2006. Age-related incident curve of hospitalized shaken baby syndrome cases: Convergent evidence for crying as a trigger to shaking. *Child Abuse and Neglect, 30,* 7–13.

Bateson, M., Nettle, D., and Roberts, G. 2006. Cues of being watched enhance cooperation in a real-world setting. *Biology Letters, 2,* 412–414.

Batsel, H. L., and Lines, A. J. 1975. Neural mechanism of the sneeze. *American Journal of Physiology, 229,* 770–776.

Bekoff, A., Stein, P. S. G., and Hamburger, V. 1975. Co-ordinated motor output in the hindlimb of the 7 day chick embryo. *Proceedings of the National Academy of Sciences, U.S., 72,* 1245–1248.

Bell, S. M., and Ainsworth, M. D. S. 1972. Infant crying and maternal responsiveness. *Child Development, 43,* 1171–1190.

Benedict, C., Frey, W. H., Schiöth, H. B., Schultes, B., Born, J., and Hallschmidt, M. 2011. Intranasal insulin as a therapeutic option in the treatment of cognitive impairments. *Experimental Gerontology, 46,* 112–115.

Bering, E. A. 1955. Choroid plexis and arterial pulsation of cerebrospinal fluid. *Archives of Neurology and Psychiatry, 73,* 165–172.

Bhutta, M. F., and Maxwell, H. 2008. Sneezing induced by sexual ideation or orgasm: An under-reported phenomenon. *Journal of the Royal Society of Medicine, 101,* 587–591.

Bigard, M.-A., Gaucher, P., and Lasalle, C. 1979. Fatal colonic explosion during colonoscopic polypectomy. *Gastroenterology, 77,* 1307–1310.

Blakemore, S.-J., Wolpert, D. M., and Firth, C. D. 1998. Central cancellation of self-produced tickle sensation. *Nature Neuroscience, 1,* 635–640.

Blank, K., Provine, R. R., and Enoch, J. M. 1975. Shift in the peak of the photopic Stiles-Crawford function with marked accommodation. *Vision Research, 15,* 499–507.

Bowlby, J. 1982. *Attachment and loss,* Vol. 1: *Attachment.* 2nd ed. New York: Basic Books.

Bramble, D. M., and D. R. Currier 1983. Running and breathing in mammals. *Science, 219,* 251–256.

Brazelton, T. B. 1962. Crying in infancy. *Pediatrics, 29,* 579–588.

Broekhuyse, R. M. 1975. The lipid composition of the aging sclera and cornea. *Ophthalmologica, 171,* 82–85.

Buhler, C., and Hetzer, H. 1928. Das ersts Verstandnis fur Ausdruck im erstn Lebengahr. *Zeitschrift fur Psychologie, 107,* 50–61.

Burnham, T. 2003. Engineering altruism: A theoretical and experimental investigation of anonymity and gift giving. *Journal of Economic Behavior and Organization, 50,* 133–144.

Burnham, T., and Hare, B. 2007. Engineering human cooperation: Does involuntary neural activation increase public goods contributions? *Human Nature, 18,* 88–108.

Campbell, M. W., Carter, J. D., Proctor, D., Eisenberg, M. L., and de Waal, F. B. M. 2009. Computer animations stimulate contagious yawning in chimpanzees. *Proceedings of the Royal Society B, 276,* 4255–4259.

Campbell, M. W., and de Waal, F. B. M. 2010. Methodological problems in the study of contagious yawning. In O. Walusinski (ed.), *The mystery of yawning in physiology and disease,* 120–127. Basel: Karger.

——. 2011. Ingroup-outgroup bias in contagious yawning by chimpanzees supports link to empathy. *PLoS One, 6,* e18283.

Carskadon, M. A. 1991. Yawning elicited by reading: Is an open mouth a sufficient stimulus. *Sleep Research, 20,* 116.

——. 1992. Yawning elicited by reading: Effects of sleepiness. *Sleep Research, 21,* 101.

Christensson, K., Cabrera, T., Christensson, E., Uvnas-Mosberg, K., and Winberg, J. 1995. Separation distress call in the human neonate in the absence of maternal body contact. *Acta Paediatrica, 84,* 468–473.

Clerf, L. H. 1947. Cough as a symptom. *Medical Clinics of North America, 31,* 1393–1399.

Collins, G. T., and Equibar, J. R. 2010. Neuropharmacology of yawning. In O. Walusinski (ed.), *The mystery of yawning in physiology and disease,* 90–106. Basel: Karger.

Criley, J. M., Blaufuss, J. H., and Kissel, G. L. 1976. Cough-induced cardiac compression: Self-administered form of cardiopulmonary resuscitation. *Journal of the American Medical Association, 236,* 1246–1250.

Criley, J. M., Niemann, J. T., Rosborough, J. P., and Hausknecht, M. 1986. Modifications of cardio-pulmonary resuscitation based on the cough. *Circulation, 74,* 42–50.

Croassin, M., Lambiase, A., Costa, N., De Gregorio, A., Sgrulletta, R., Sacchetti, M., Aloe, L., and Bonini, S. 2005. Efficacy of topical nerve growth factor treatment in dogs affected by dry eye. *Graefe's Archive for Clinical and Experimental Ophthalmology, 243,* 151–155.

Darwin, C. 1965 [1872]. *The expression of emotions in man and animals.* Chicago: University of Chicago Press.

Davidson, S., and Giesler, G. J. 2010. The multiple pathways for itch and their interactions with pain. *Trends in Neuroscience, 33,* 550–558.

Davidson, S., Zhang, X., Khasabov, S. G., Simone, D. A., and Giesler, G. J. Jr. 2009. Relief of itch by scratching: State-dependent inhibition of primate spinothalamic tract neurons. *Nature Neuroscience, 12,* 544–546.

Davila-Ross, M., Allcock, B., Thomas, C., and Bard, K. A. 2011. Aping expressions? Chimpanzees produce distinct laugh types when responding to laughter of others. *Emotion, 11,* 1113–1120.

Davila-Ross, M., Owren, M. J., and Zimmermann, E. 2009. Reconstructing the evolution of laughter in great apes and humans. *Current Biology, 19,* 1106–1111.

Davis, S. A. 1983. Cough-CPR and a new theory of blood flow. *Critical Care Nurse, 3,* 42–46.

Decety, J., and Jackson, P. L. 2004. The functional architecture of human empathy. *Behavioral and Neuroscience Reviews, 3,* 71–100.

Deputte, B. L. 1994. Ethological study of yawning in primates. I. Quantitative analysis and study of causation in two species of Old World monkeys (*Cercocebus albigena* and *Macaca fascicularis*). *Ethology, 98,* 221–245.

Dethier, V. G. 1963. *To know a fly.* San Francisco: Holden-Day.

de Vries, J. I. P., Visser, G. H., and Prechtl, H. F. 1982. The emergence of fetal behaviour. I. Qualitative aspects. *Early Human Development, 7,* 301–322.

——. 1985. The emergence of fetal behaviour. II. Quantitative aspects. *Early Human Development, 12,* 99–120.

Dondi, M., Simion, F., and Caltran, G. 1999. Can newborns discriminate between their own cry and the cry of another newborn infant? *Developmental Psychology, 35,* 418–426.

Donshik, P. C. 1988. Allergic conjunctivitis. *International Ophthalmology Clinics, 28,* 294–301.

Drachman, D. B., and Coulumbre, A. J. 1962. Experimental clubfoot and arthrogryposis multiplex congenital. *Lancet, 2,* 523–526.

Drachman, D. B., and Sokoloff, L. 1966. The role of movement in embryonic joint development. *Developmental Biology, 14,* 401–420.

du Boulay, G., O'Connell, J., Currie, J., Bostick, T., and Verity, P. 1972. Further investigations of pulsatile movements in the cerebrospinal fluid pathways. *Acta Radiologica, 13,* 496–523.

Duman, R. S., and Monteggia, L. M. 2006. A neurotrophic model for stress-related mood disorders. *Biological Psychiatry, 59,* 1116–1127.

Dumpert, V. 1929. Uber die Bedeutung des Singultus. *Deutsche Zeitschrift fur Nervenheilkunde, 110,* 106–116.

Dunbar, R. I. M., Baron, R., Frangou, A., Pearce, E., van Leeuwen, E. J. C., Stow, J., Partridge, G., MacDonald, I., Barra, V., and van Vugt, M. In press. Social laughter is correlated with an elevated pain threshold. *Proceedings of the Royal Society B.*

Eibl-Eibesfeldt, I. 1973. The expressive behavior of the deaf-blind born. In M. van Cranach and I. Vine (eds.), *Social communication and movement*, 163–194. London: Academic Press.

——. 1975. *Ethology*. 2nd ed. New York: Holt, Rinehart and Winston.

Ekman, P., Friesen, W. V., and Ellsworth, P. 1972. *Emotion in the human face*. New York: Pergamon.

Emery, N. J. 2000. The eyes have it: The neuroethology, function and evolution of social gaze. *Neuroscience and Biobehavioral Review, 24,* 581–604.

Etcoff, N. 1999. *Survival of the prettiest*. New York: Doubleday.

Everett, H. C. 1964. Sneezing in response to light. *Neurology, 14,* 483–490.

Fesmire, F. M. 1988. Termination of intractable hiccup with digital rectal massage. *Annals of Emergency Medicine, 17,* 872.

Field, T., Diego, M., Hernandez-Reif, M., and Fernandez, M. 2007. Depressed mothers' newborns show less discrimination of other newborns' cry sounds. *Infant Behavior and Development, 30,* 431–435.

Firth, R. 2004 [1936]. *We, the Tikopia: A sociological study of kindship in primitive Polynesia*. London: Routledge.

Firth, S. I., Wang, C. T., and Feller, M. B. 2005. Retinal waves: Mechanisms and function in visual system development. *Cell Calcium, 37,* 425–432.

Fisher, C. M. 1967. Protracted hiccup—a male malady. *Transactions of the American Neurological Association, 92,* 231–233.

Freedman, D. G. 1964. Smiling in blind infants and the issue of innate vs. acquired. *Journal of Child Psychology and Psychiatry, 5,* 171–184.

Frey, W. H. 1985. *Crying: The mystery of tears*. Minneapolis, MN: Winston Press.

Fridlund, A. J. 1994. *Human facial expression*. San Diego, CA: Academic Press.

Friedman, H. S., Tucker, J. S., Tomlinson-Keasey, C., Schwartz, J. E., Wingard, D. L, and Criqui, M. H. 1993. Does childhood personality predict longevity? *Journal of Personality and Social Psychology, 65,* 176–185.

Friedman, N. L. 1996. Hiccups: A treatment review. *Pharmacotherapy, 16,* 986–995.

Fossey, D. 1972. Vocalizations in the mountain gorilla *(Gorilla beringei)*. *Animal Behaviour, 20,* 36–53.

Fouts, R. 1997. *Next of kin*. New York: William Morrow.

Fuller, G. N. 1990. Hiccups and human purpose. *Nature, 343,* 420.

Gallia, L. J., and Roscoe, G. 1981. Intractable sneezing. *Transactions of the Pennsylvania Academy of Ophthalmology and Otolaryngology, 34,* 164–168.

Gallup, A. C. 2010. A thermoregulatory behavior. In O. Walusinski (ed.), *The mystery of yawning in physiology and disease,* 84–89. Basel: Karger.

Gallup, A. C., and Gallup, G. G. 2007. Yawning as a brain cooling mechanism: Nasal breathing and forehead cooling diminish the incidence of contagious yawning. *Evolutionary Psychology, 5,* 92–101.

Gallup, G. G. 1970. Chimpanzees: Self-recognition. *Science, 167,* 86–87.

Gangestad, S. W., and Thornhill, R. 1999. Individual differences in developmental precision and fluctuating asymmetry: A model and its implications. *Journal of Evolutionary Biology, 12,* 402–416.

Gawande, A. 2008. The itch. *New Yorker,* June 30, 58–65.

Geangu, E., Benga, O., Stahl, D., and Triano, T. 2010. Contagious crying beyond the first days of life. *Infant Behavior and Development, 33,* 279–288.

Gekowski, M. J., Rovee-Collier, C. K., and Carulli-Rabinowitz, V. 1983. A longitudinal analysis of inhibition of infant distress: The origins of social expectations? *Infant Behavior and Development, 6,* 341–351.

Gelstein, S., Yeshurum, Y., Rosenkrantz, L., Shushan, S., Frumin, I., Roth, Y., and Sobel, N. 2011. Human tears contain a chemosignal. *Science, 331,* 226–230.

Gervais, M., and Wilson, D. S. 2005. The evolution and function of laughter and humor: A synthetic approach. *Quarterly Review of Biology, 80,* 395–430.

Gessell, A., and Amatruda, C. S. 1945. *The embryology of behavior.* New York: Harper.

Gessell, A., and Ilg, F. L. 1943. *Infant and child in the culture of today: The guidance of development in home and nursery school.* New York: Harper Collins.

Gieler, U., Niemeier, V., Brosig, B., and Kupfer, J. 2002. Psychosomatic aspects of pruritus. *Dermatology and Psychosomatics, 3,* 6–13.

Giganti, F., and Esposito Ziello, M. 2009. Contagious and spontaneous yawning in autistic and typically developing children. *Current Psychology Letters, 25,* 1–11.

Giganti, F., Hayes, M., Cioni, G., and Salzarulo, P. 2007. Yawning frequency and distribution in preterm and near term infants assessed through 24-h recordings. *Infant Behavior Development, 30,* 641–647.

Girsky, M. J., and Criley, J. M. 2006. Cough cardiopulmonary resuscitation revisited, *14,* 530–531.

Golumb, G. 1990. Hiccup for hiccups. *Nature, 345,* 774.

Gomez, C., David, V., Peet, N. M., Vico, L., Chenu, C., Malaval, L., and Skerry, T. M. 2007. Absence of mechanical loading in utero influences bone mass and architecture but not innervation in Myod-Myf5-deficient mice. *Journal of Anatomy, 210,* 259–271.

Goodall, J. 1986. *The chimpanzees of Gombe: Patterns of behavior.* Cambridge, MA: Harvard University Press.

Gottlieb, G. 1973. Dedication to W. Preyer 1841–1897. In G. Gottlieb (ed.), *Behavioral embryology*, xv–xix. New York: Academic Press.

Grammer, K. 1990. Strangers meet: Laughter and non-verbal signs of interest in opposite-sex encounters. *Journal of Nonverbal Behavior, 14,* 209–236.

Grammer, K., and Eibl-Eibesfeldt, I. 1990. The ritualization of laughter. In W. A. Koch (ed.), *Naturlichtkeit der Sprache und der Kulture. Bochumer Beitrage zur Semiotik,* 192–214. Bochum, Germany: Brockmeyer.

Grandin, T. 1995. *Thinking in pictures*. New York: Doubleday.

Gruber, J., Johnson, S. L., Oveis, C., and Keltner, D. 2008. Risk for mania and positive emotional responding: Too much of a good thing? *Emotion, 8,* 23–33.

Gruber, J., Mauss, I. B., and Tamir, M. 2011. A dark side of happiness? How, when, and why happiness is not always good. *Perspectives on Psychological Science, 6,* 222–233.

Guggisberg, A. G., Mathis, J., and Hess, C. W. 2010. Interplay between yawning and vigilance: A review of the experimental evidence. In O. Walusinski (ed.), *The mystery of yawning in physiology and disease,* 47–54. Basel: Karger.

Guggisberg, A. G., Mathis, J., Schnider, A., and Hess, C. W. 2010. Why do we yawn? *Neuroscience and Biobehavioral Reviews, 34,* 1267–1276.

Gutmann, B. 1926. *Das recht der Dschagga*. Munich: Beck.

Haker, H., and Rossler, W. 2009. Empathy in schizophrenia: Impaired resonance. *European Archives of Psychiatry and Clinical Neuroscience, 259,* 352–361.

Haley, K. J., and Fessler, D. M. T. 2005. Nobody's watching? Subtle cues affect generosity in an anonymous economic game. *Evolution and Human Behavior, 26,* 245–256.

Hamburger, V. 1963. Some aspects of the embryology of behavior. *Quarterly Review of Biology, 38,* 342–365.

Hamburger, V., and Oppenheim, R. W. 1982. Naturally occurring neuronal death in vertebrates. *Neuroscience Commentaries, 1,* 39–55.

Hamburger, V., Wenger, E., and Oppenheim, R. W. 1966. Motility in the chick embryo in the absence of sensory input. *Journal of Experimental Embryology, 162,* 133–160.

Handwerker, H. O., Forster, C., and Kirchoff, C. 1991. Discharge patterns of human C-fibers induced by itching and burning stimuli. *Journal of Neurophysiology, 66,* 307–315.

Harr, A., Gilbert, V., and Phillips, K. 2009. Do dogs *(Canis familiaris)* show contagious yawning? *Animal Cognition, 12,* 833–837.

Harris, C. R. 1999. The mystery of ticklish laughter. *American Scientist, 87,* 344–351.

Hatfield, E., Cacciopo, J., and Rapson, R. 1994. *Emotional contagion*. New York: Cambridge University Press.

Haxby, J. V., Hoffman, E. A., and Gobbini, J. 2000. The distributed human neural system for face perception. *Trends in Cognitive Science, 4,* 223–233.

Heath, C. 1989. Pain talk: The expression of suffering in the medical consultation. *Social Psychology Quarterly, 52,* 113–125.

Hefez, A. 1985. The role of the press and the medical community in the epidemic of "mysterious gas poisoning" in the Jordan West Bank. *American Journal of Psychiatry, 142,* 833–837.

Helt, M. S., Eigsti, I.-M., Snyder, P. J., and Fein, D. A. 2010. Contagious yawning in autistic and typical development. *Child Development, 81,* 1620–1631.

Hemplemann, C. F. 2007. The laughter of the Tanganyikan "laughter epidemic." *Humor, 20,* 49–71.

Hersch, M. 2000. Loss of ability to sneeze in lateral medullary syndrome. *Neurology, 54,* 520–521.

Hess, E. H. 1975. *The tell-tale eye: How your eyes reveal hidden thoughts and emotions.* New York: Van Nostrand Reinhold.

Heusner, A. P. 1946. Yawning and associated phenomena. *Physiological Review, 25,* 156–168.

Hippocrates. 1981. *Works.* Trans. W. H. S. Jones. Cambridge, MA: Harvard University Press.

Hoebel, B. G., Rada, P. V., Mark, G. P., and Pothos, E. 1999. Neural systems for reinforcement and inhibition behavior: Relevance to eating, addition, and depression. In D. Kahneman, E. Diener, and N. Schwart (eds.), *Well-being: Foundations of hedonic psychology,* 560–574. New York: Russell Sage Foundation.

Hoffman, M. L. 2000. *Empathy and moral development: Implications for caring and justice.* Cambridge: Cambridge University Press.

Hopkins, B. 2000. Development of crying in normal infants: Method, theory and some speculations. In R. G. Barr, B. Hopkins, and J. A. Green (eds.), *Crying as a sign, a symptom, and a signal,* 176–209. London: MacKeith Press.

Hunziker, U. A., and Barr, R. G. 1986. Increased carrying reduces infant crying: A randomized control trial. *Pediatrics, 77,* 641–648.

Hurley, M. M., Dennett, D. C., and Adams, R. B. 2011. *Insider jokes: Using humor to reverse-engineer the mind.* Cambridge, MA: MIT Press.

Iacoboni, M. 2009. Imitation, empathy, and mirror neurons. *Annual Review of Psychology, 60,* 653–670.

Irwin, R. S., Curley, F. J., and French, C. 1990. Chronic cough. *American Review of Respiratory Diseases, 141,* 640–647.

Isenberg, S. J., Apt, L., McCarthy, J., Copper, L., Lim, and Del Signore, M. 1998. Developing of tearing in preterm and term neonates. *Archives of Ophthalmology, 116,* 773–776.

Jago, R. H. 1970. Arthrogryposis following treatment of maternal tetanus with muscle relaxants. *Archives of Diseases of Children, 45,* 277–279.

Johanek, L. M., Meyer, R. A., Hartke, T., Hobelmann, J. G., Maine, D. N., La Motte, R. H., and Ringkamp, M. 2007. Psychophysical and physiological evidence for parallel afferent pathways mediating the sensation of itch. *Journal of Neuroscience, 27,* 7490–7497.

Johnston, V. S. 2006. Mate choice decisions: The role of facial beauty. *Trends in Cognitive Science, 10,* 9–13.

Joly-Mascheroni, R. M., Senju, A., and Shepherd, A. J. 2008. Dogs catch human yawns. *Biology Letters, 4,* 446–448.

Kalat, J. W. 2009. Novocain anecdote. *Biological Psychology,* 10th ed., 208. Belmont, CA: Wadsworth.

Kerr, A., and Eich, R. H. 1961. Cerebral concussion as a cause of cough syncope. *Archives of Internal Medicine, 108,* 248–252.

Kiln, A., Jones, W., Schultz, R., Volkmar, F., and Cohen, D. 2002. Visual fixation patterns during viewing of naturalistic social situations as predictors of social competence in individuals with autism. *Archives of General Psychiatry, 59,* 809–816.

Klesius, M. 2009. Sick in space: It's not just a problem for astronauts anymore. http://www.airspacemag.com.com/space-exploration/Sick-in-space.html?c=yandpage=1.

Kobayashi, H., and Kohshima, S. 2001. Unique morphology of the human eye and its adaptive meaning: Comparative studies on external morphology of the primate eye. *Journal of Human Evolution, 40,* 419–435.

Kraemer, D. L., and Hastrup, J. L. 1988. Crying in adults: Self-control and autonomic correlates. *Journal of Social and Clinical Psychology, 6,* 53–68.

Lambiase, A., Rama, P., Bonini, S., Caprioglio, G., and Aloe, L. 1998. Topical treatment with nerve growth factor for corneal neurotropic ulcers. *New England Journal of Medicine, 338,* 1174–1180.

Landmesser, L., and O'Donovan, M. J. 1984. Activation patterns of embryonic chick hind limb muscles recorded *in ovo* and in an isolated spinal cord preparation. *Journal of Physiology, 347,* 189–204.

Lang, I. M., and Sarna, S. K. 1989. Motor and myoelectric activity associated with vomiting, regurgitation, and nausea. In J. D. Wood (ed.), *Handbook of physiology: Gastrointestinal motility and circulation,* 1179–1198. Bethesda, MD: MPT Press Ltd.

Langlois, J. H., Kalakanis, L., Rubenstein, A. J., Larson, A., Hallamm, M., and Smoot, M. 2000. Maxims or myths of beauty? A meta-analytic and theoretical review. *Psychological Bulletin, 126,* 390–423.

Launois, S., Bizec, J. L., Whitelaw, W. A., Cabane, J., and Derenne, J. P. 1993. Hiccups in adults: An overview. *European Respiratory Journal, 6,* 563–575.

Lehmann, H. E. 1979. Yawning, a homeostatic reflex and its physiological significance. *Bulletin of the Menninger Clinic, 43,* 123–136.

Leibowitz, H. M. 2000. The red eye. *New England Journal of Medicine, 343,* 345–351.

Levi-Montalcini, R. 1988. *In praise of imperfection: My life and work.* New York: Basic Books.

Levine, R. J. 1977. Epidemic faintness and syncope in a school marching band. *Journal of the American Medical Association, 238,* 2373–2376.

Levitt, M. D. 1984. Gastrointestinal gas and abdominal symptoms. *Practical Gastroenterology, 7,* 4–14.

Lewis, J. H. 1985. Hiccups: Causes and cures. *Journal of Gastroenterology, 7,* 539–552.

Lieberman, A., and Benson, F. 1977. Control of emotional expression in pseudobulbar palsy: A personal experience. *Archives of Neurology, 34,* 717–719.

Little, A. C., Apicella, C. L., and Marlowe, F. W. 2007. Preferences for symmetry in human faces in two cultures: Data from the UK and the Hadza, an isolated group of hunter gathers. *Proceedings of the Royal Society B, 274,* 3113–3117.

Lockey, M. P., Poots, G., and Williams, B. 1975. Theoretical aspects of the attenuation of pressure pulses within cerebrospinal-fluid pathways. *Medical and Biological Engineering, 14,* 861–869.

Luchinger, A. B., Hadders-Algra, M., Van Kan, C. M., and de Vries, J. I. P. 2008. Fetal onset of general movements. *Pediatric Research, 63,* 191–195.

Lutz, T. 1999. *Crying: The natural and cultural history of tears.* New York: Norton.

MacLarnon, A. M., and Hewitt, G. P. 1999. The evolution of human speech: The role of enhanced breathing control. *American Journal of Physical Anthropology, 109,* 341–363.

Makagon, M. M., Funayama, E. S., and Owren, M. J. 2008. An acoustic analysis of laughter produced by congenitally deaf and normally hearing college students. *Journal of the Acoustical Society of America, 124,* 472–483.

Martin, G., and Clark, R. 1982. Distress crying in neonates: Species and peer specificity. *Developmental Psychology, 18,* 3–9.

Martin, R. A. 2007. *The psychology of humor.* San Diego, CA: Academic Press.

Masson, J. M., and McCarthy, S. 1995. *When elephants weep: The emotional lives of animals.* New York: Delacorte Press.

Matsusaka, T. 2004. When does play panting occur during social play in wild chimpanzees? *Primates, 45,* 221–229.

Mayo, C. W. 1932. Hiccup. *Surgery, Gynecology, and Obstetrics, 55,* 700–708.

McFarland, D. H. 2001. Respiratory markers of conversational interaction. *Journal of Speech, Language, and Hearing, 44,* 128–143.

McLane, N. J., and Carroll, D. M. 1986. Ocular manifestations of drug abuse. *Survey of Ophthalmology, 30,* 298–313.

Mead, M., and Newton, N. 1967. Cultural patterning in perinatal behavior. In S. A. Richardson and A. F. Guttmacher (eds.), *Childbearing: Its social and psychological aspects,* 142–244. Baltimore, MD: Williams and Wilkins.

Meenakshisundaram, R., Thirumalaikolundusubramanian, P., Walusinski, O., Muthusundari, A., and Sweni, S. 2010. Associated movements in hemiplegic limbs during yawning. In O. Walusinski (ed.), *The mystery of yawning in physiology and disease,* 134–139. Basel: Karger.

Miller, M., and Fry, W. F. 2009. The effect of mirthful laughter on the human cardiovascular system. *Medical Hypotheses, 73,* 636–639.

Miller, M. E., Higginbottom, M., and Smith, D. W. 1981. Short umbilical cord: Its origin and relevance. *Pediatrics, 67,* 618–621.

Moessinger, A. C. 1988. Morphological consequences of depressed or impaired fetal activity. In W. P. Smotherman and S. R. Robinson (eds.), *Behavior of the fetus,* 163–173. Caldwell, NJ: Telford Press.

Moessinger, A. C., Blanc, W. A., Marone, P. A., and Polsen, D. C. 1982. Umbilical cord length as an index of fetal activity: Experimental study and clinical implications. *Pediatrics Research, 16,* 109–112.

Moffatt, M. E. K. 1982. Epidemic hysteria in a Montreal train station. *Pediatrics, 70,* 308–310.

Montaigne, M. E. de. 1958 [1595]. *The complete essays of Montaigne.* Trans. D. M. Frame. Stanford, CA: Stanford University Press.

Moore, J. E. 1942. Some psychological aspects of yawning. *Journal of General Psychology, 27,* 289–294.

Mueller, J. B., and McStay, C. M. 2008. Ocular infection and inflammation. *Emergency Medical Clinics of North America, 26,* 57–72.

Mulley, G. 1982. Associated reactions in the hemiplegic arm. *Scandinavian Journal of Rehabilitation Medicine, 14,* 117–120.

Murphy, P. J., Lau, J. S. C., Sim, M. M. L., and Woods, R. L. 2007. How red is a white eye? Clinical grading of normal conjunctival hyperemia. *Eye, 21,* 633–638.

Murray, N., and Bierer, J. 1951. Prolonged sneezing. *Psychomatic Medicine, 13,* 56–58.

Nahab, F. B. 2010. Exploring yawning with neuroimaging. In O. Walusinski (ed.), *The mystery of yawning in physiology and disease,* 128–133. Basel: Karger.

Nahab, F. B., Hattori, N., Saad, Z. S., and Hallett, M. 2009. Contagious yawning and the frontal lobe: An fMRI study. *Human Brain Mapping, 30,* 1744–1751.

Nemery, B., Fischer, B., Boogaerts, M., Lison, D., and Willems, J. 2002. The Coca- Cola incident in Belgium, June 1999. *Food and Chemical Toxicology, 40,* 1657–1667.

Newsom-Davis, J. 1970. An experimental study of hiccup. *Brain, 93,* 851–872.

Nicholson, C. 2010. The humor gap. *Scientific American Mind,* May/June 2010, 38–45.

Niemeier, V., Kupfer, J., and Gieler, U. 2000. Observations during itch-inducing lecture. *Dermatology and Psychosomatics, 1,* 15–18.

Nohain, J., and Caradec, F. 1967. *Le Pétomane.* Los Angeles: Sherbourne Press.

Nonaka, S., Unno, T., Ohta, Y., and Mori, S. 1990. Sneeze-evoking region within the brainstem. *Brain Research, 511,* 265–270.

Norn, M. 1985. Pigment spots related to scleral emissaries in Eskimos, Mongols, and Caucasians. *Acta Ophthalmologica, 63,* 236–241.

Odeh, M., Bassan, H., and Oliven, A. 1990. Termination of intractable hiccups with digital rectal massage. *Journal of Internal Medicine, 227,* 145–146.

O'Hara, S. J., and Reeve, A. V. 2011. A test of the yawning contagion and emotional connectedness hypothesis in dogs, *Canis familiaris. Animal Behaviour, 81,* 335–340.

Oleshansky, M. A., and Labbate, L. A. 1996. Inability to cry during SRI treatment. *Journal of Clinical Psychiatry, 57,* 593.

Oppenheim, R. W. 1984. Ontogenetic adaptations in neural development: Toward a more "ecological" developmental psychobiology. In H. F. R. Prechtl (ed.), *Continuity of neural functions from prenatal to postnatal life,* 16–30. Philadelphia: Lippincott.

Oppenheim, R. W., Caldero, J., Esquerda, J., and Gould, T. W. 2001. Target-independent programmed cell death in the developing nervous system. In A. F. Kalverboer and A. Gramsbergen (eds.), *Handbook of brain and behaviour in human development,* 343–408. Dordrecht: Kluwer Academic Publishers.

Oppenheim, R. W., and Nunez, R. 1982. Electrical stimulation of hindlimb increases neuronal cell death in chick embryos. *Nature, 295,* 57–59.

Osborne, A. C., Lamb, K. J., Lewthwaite, J. C., Dowthwaite, G. P., and Pitsillides, A. A. 2002. Short-term rigid and flaccid paralyses diminish growth of embryonic chick limbs and abrogate joint cavity formation but differentially preserve precavitated joints. *Journal of Musculoskeletal and Neuronal Interaction, 2,* 448–456.

Ostwald, P. 1972. The sounds of infancy. *Developmental Medicine and Child Neurology, 14,* 350–361.

Owen, C. G., Newsom, R. S. B., Rudnicka, A. R., Ellis, T. J., and Woodward, E. G. 2005. Vascular response of the bulbar conjunctiva to diabetes and elevated blood pressure. *Ophthalmology, 112,* 1801–1808.

Palagi, E., Leone, A., Mancini, G., and Ferrari, P. F. 2009. Contagious yawning in gelada baboons as a possible expression of empathy. *Proceedings of the National Academy of Sciences, U.S., 106,* 19262–19267.

Panksepp, J. 2007. Neuroevolutionary sources of laughter and social joy: Modeling primal human laughter in laboratory rats. *Behavioral Brain Research, 182,* 231–244.

Panksepp, J., and Burgdorf, J. 1999. Laughing rats? Playful tickling arouses high frequency ultrasound chirping in young rodents. In S. Hameroff, D. Chalmers, and A. Kaziak (eds.), *Toward a science of consciousness III,* 231–244. Cambridge, MA: MIT Press.

——. 2003. "Laughing" rats and the evolutionary antecedents of human joy? *Physiology and Behavior, 79,* 533–547.

Papavramidou, N., Fee, E., and Christopoulou-Aletra, H. 2007. Jaundice in the *Hippocratic corpus. Journal of Gastrointestinal Surgery, 11,* 1728–1731.

Papoiu, A. S. P., Wang, H., Coghill, R. C., Chan, Y.-H., and Yosipovitch, G. 2011. Contagious itch in humans. A study of visual "transmission" of itch in atopic dermatitis and healthy subjects. *British Journal of Dermatology,* in press.

Parvizi, J., Anderson, S. W., Martin, C. O., Damasio, H., and Damasio, A. R. 2001. Pathological laughter and crying: A link to the cerebellum. *Brain, 124,* 1708–1719.

Paton, D. 1961. The conjunctival sign of sickle cell disease. *Archives of Ophthalmology, 66,* 90–94.

Paukner, A., and Anderson, J. R. 2006. Video-induced yawning in stumptail macaques *(Macaca arctoides). Biology Letters, 2,* 36–38.

Peleg, R., and Peleg, A. 2000. Case report: Sexual intercourse as potential treatment for intractable hiccups. *Canadian Family Physician, 46,* 1631–1632.

Pennebaker, J. W. 1980. Perceptual and environmental determinants of coughing. *Basic and Applied Social Psychology, 1,* 83–91.

Pentland, A. 2008. *Honest signals: How they shape our world.* Cambridge, MA: MIT Press.

Perrett, D. I., and Mistlin, A. J. 1991. Perception of facial characteristics by monkeys. In W. C. Stebbins and M. A. Berkley (eds.), *Comparative perception,* 187–215. New York: John Wiley.

Petelenz, T., Iwinski, J., Chelbowszyx, J., Czyx, Z., Flak, Z., Fiutowski, L., Zaoski, K., Petelenz, T., and Zeman, S. 1998. Self-administered cough cardiopulmonary resuscitation (c-CPR) inpatients threatened by MAS events of cardiovascular origin. *Wiadomosci lekarski, 51,* 326–336.

Pillai, M., and James, D. 1990. Hiccups and breathing in human fetuses. *Archives of Disease in Childhood, 65,* 1072–1075.

Piontelli, A. 2010. *Development of normal fetal movements.* Milan, Italy: Springer.

Platek, S. M. 2010. Yawn, yawn, yawn, yawn, yawn, yawn! The social, evolutionary and neuroscientific facets of contagious yawning. In O. Walusinski (ed.), *The mystery of yawning in physiology and disease,* 107–112. Basel: Karger.

Platek, S. M., Critton, S. R., Myers, T. E., and Gallup, G. G. 2003. Contagious yawning: The role of self-awareness and mental state attribution. *Cognitive Brain Research, 17,* 223–227.

Platek, S. M., Mohamed, F. B., and Gallup, G. G. 2005. Contagious yawning and the brain. *Cognitive Brain Research, 23,* 448–452.

Pliny. 1951. *Natural History.* Trans. W. H. S. Jones. London: Heinemann.

Plooij, F. 1979. How wild chimpanzee babies trigger the onset of mother-infant play—and what the mother makes of it. In M. Bullowa (ed.), *Before speech: The beginning of interpersonal communications,* 223–243. Cambridge: Cambridge University Press.

Plum, F. 1970. General discussion. In R. Porter (ed.), *Breathing: Hering-Breuer centenary symposium. A Ciba foundation symposium,* 203. London: Churchill.

Pittman, R., and Oppenheim, R. W. 1978. Neuromuscular blockade increases motoneuron survival during normal cell death in the chick embryo. *Nature, 271,* 364–366.

Provine, R. R. 1971. Embryonic spinal cord: Synchrony and spatial distribution of polyneuronal burst discharges. *Brain Research, 29,* 155–158.

——. 1972. Ontogeny of bioelectric activity in the spinal cord of the chick embryo and its behavioral implications. *Brain Research, 41,* 365–378.

——. 1973. Neurophysiological aspects of behavior development in the chick embryo. In G. Gottlieb (ed.), *Behavioral embryology,* 77–102. New York: Academic Press.

——. 1980. Development of between-limb movement synchronization in the chick embryo. *Developmental Psychobiology, 13,* 151–163.

——. 1986. Yawning as a stereotyped action pattern and releasing stimulus. *Ethology, 72,* 109–122.

——. 1988. On the uniqueness of embryos and the difference it makes. In W. P. Smotherman and S. R. Robinson (eds.), *Behavior of the fetus,* 35–46. Caldwell, NJ: Telford Press.

——. 1989a. Contagious yawning and infant imitation. *Bulletin of the Psychonomic Society, 27,* 125–126.

——. 1989b. Faces as releasers of contagious yawning: An approach to face detection using normal human subjects. *Bulletin of the Psychonomic Society, 27,* 211–214.

——. 1989c. Yawning and simulation science. *Simulation, 53,* 193–194.

——. 1992. Contagious laughter: Laughter is a sufficient stimulus for laughs and smiles. *Bulletin of the Psychonomic Society, 30,* 1–4.

——. 1993. Laughter punctuates speech: Linguistic, social and gender contexts of laughter. *Ethology, 95,* 291–298.

——. 1996a. Laughter. *American Scientist, 84,* 38–45.

——. 1996b. Contagious yawning and laughter: Significance for sensory feature detection, motor pattern generation, imitation, and the evolution of social behavior. In C. M. Heyes and B. G. Galef (eds.), *Social learning in animals: The roots of culture,* 179–208. San Diego, CA: Academic Press.

——. 1997a. Bipedalism and speech evolution. *Society for Neuroscience Abstracts, 23,* part 2, 2134.

——. 1997b. Yawns, laughs, smiles, tickles, and talking: Naturalistic and laboratory studies of facial action and communication. In J. A. Russell and J. M. Fernandez-Dols (eds.), *New directions in the study of facial expression,* 158–175. Cambridge: Cambridge University Press.

——. 1999. Stand up and talk: Bipedalism and speech evolution. *Society for Neuroscience Abstracts, 25,* part 2, 2170.

——. 2000. *Laughter: A scientific investigation.* New York: Viking.

——. 2004. Laughing, tickling, and the evolution of speech and self. *Current Directions in Psychological Science, 13,* 25–218.

——. 2005. Yawning. *American Scientist, 93,* 532–539.

——. 2011. Emotional tears and NGF: A biographical appreciation and research beginning. *Archives Italiennes de Biologie, 149,* 269–274.

Provine, R. R., and Bard, K. A. 1994. Laughter in chimpanzees and humans: A comparison. *Society for Neuroscience Abstracts, 20,* part 1, 367.

——. 1995. Why chimps can't talk: The laugh probe. *Society for Neuroscience Abstracts, 21,* part 1, 456.

Provine, R. R., Cabrera, M. O., Brocato, N. W., and Krosnowski, K. A. 2011. When the whites of eyes are red: A uniquely human cue. *Ethology, 117,* 1–5.

Provine, R. R., and Emmorey, K. 2006. Laughter among deaf signers. *Journal of Deaf Studies and Deaf Education, 11,* 403–409.

Provine, R. R., and Enoch, J. M. 1975. On ocular accommodation. *Perception and Psychophysics, 17,* 209–212.

Provine, R. R., and Fischer, K. R. 1989. Laughing, smiling and talking: Relation to sleeping and social context in humans. *Ethology, 83,* 295–305.

Provine, R. R. and Hamernik, H. B. 1986. Yawning: Effects of stimulus interest. *Bulletin of the Psychonomic Society, 24,* 437–438.

Provine, R. R., Hamernik, H. B., and Curchack, B. C. 1987. Yawning: Relation to sleeping and stretching in humans. *Ethology, 76,* 152–160.

Provine, R. R., Krosnowski, K. A., and Brocato, N. W. 2009. Tearing: Breakthough in human emotional signaling. *Evolutionary Psychology, 7,* 52–56.

Provine, R. R., and Rogers, L. 1977. Development of spinal cord bioelectric activity in spinal chick embryos and its behavioral implications. *Journal of Neurobiology, 8,* 217–228.

Provine, R. R., Sharma, S. C., Sandel, T. T., and Hamburger, V. 1970. Electrical activity in the spinal cord of the chick embryo in situ. *Proceedings of the National Academy of Sciences, U.S., 65,* 508–515.

Provine, R. R., Spencer, R. J., and Mandell, D. L. 2005. Emoticons punctuate website text messages. *Journal of Language and Social Psychology, 26,* 299–307.

Provine, R. R., Tate, B. C., and Geldmacher, L. L. 1987. Yawning: No effect of 3–5% CO_2, 100% O_2, and exercise. *Behavioral and Neural Biology, 48,* 382–393.

Provine, R. R., and Westerman, J. 1979. Crossing the midline: Limits of eye-hand behavior. *Child Development, 50,* 437–441.

Provine, R. R., and Yong, Y. L. 1991. Laughter: A stereotyped human vocalization. *Ethology, 89,* 115–124.

Ramachandran, V. S., and Blakeslee, S. 1998. *Phantoms in the brain.* New York: William Morrow.

Ramachandran, V. S., and Oberman, L. M. 2006. Broken mirrors: A theory of autism. *Scientific American, 295,* 62–69.

Rankin, A. M., and Philip, P. J. 1963. An epidemic of laughter in the Bukoba district of Tanganyika. *Central African Journal of Medicine, 9,* 167–170.

Raum, O. F. 1939. Female initiation among the Chagga. *American Anthropologist, 41,* 554–565.

Rhodes, G. 2006. The evolutionary psychology of facial beauty. *Annual Review of Psychology, 57,* 199–226.

Ripley, K. L., and Provine, R. R. 1972. Neural correlates of embryonic motility in the chick. *Brain Research, 45,* 127–134.

Rizzolatti, G., and Fabbri-Destro, M. 2010. Mirror neurons: From discovery to autism. *Experimental Brain Research, 200,* 223–237.

Robinson, P., Szewczyk, M., Haddy, L., Jones, P., and Harvey, W. 1984. Outbreak of itching and rash: Epidemic by hysteria in an elementary school. *Archives of Internal Medicine, 144,* 1959–1962.

Roche, S. P., and Kobos, R. 2004. Jaundice in the adult patient. *American Family Physician, 69,* 299–304.

Rockney, R. M., and Lemke, T. 1992. Casualties from a junior-senior high school during the Persian Gulf War: Toxic poisoning or mass hysteria? *Developmental and Behavioral Pediatrics, 13,* 339–342.

Rosenberg, M. 2009. Cause of Afghan girls' illness a mystery. *Wall Street Journal,* Tuesday, May 19, A10.

Rothbart, M. K. 1973. Laughter in young children. *Psychological Bulletin, 80,* 247–256.

Rottenberg, J., Bylsma, L. M., and Vingerhoets, A. J. J. M. 2008. Is crying beneficial? *Current Directions in Psychological Science, 17,* 400–404.

Rozin, P., and Fallon, A. E. 1987. A perspective on disgust. *Psychological Review, 94,* 23–41.

Rozin, P., Haidt, J., and McCauley, C. R. 2000. Disgust. In M. Lewis and M. J. Haviland (eds.), *Handbook of emotions,* 2nd ed., 637–653. New York: Guilford.

Rozin, P., and Kalat, J. 1971. Specific hungers and poison avoidance as adaptive specializations. *Psychological Review, 78,* 459–486.

Rozin, P., Millman, L., and Nemeroff, C. 1986. Operation of the laws of sympathetic magic in disgust and other domains. *Journal of Personality and Social Psychology, 50,* 703–712.

Sadenand, V., Kelly, M., Varughese, G., and Forney, D. R. 2005. Sudden quadriplegia after acute cervical disk herniation. *Canadian Journal of Neurological Science, 32,* 356–358.

Sagi, A., and Hoffman, M. L. 1976. Empathetic distress in the newborn. *Developmental Psychology, 12,* 175–176.

Salmons, S., and Streter, F. A. 1976. Significance of impulse activity in the transformation of skeletal muscle type. *Nature, 263,* 30–34.

Sarkies, M. 1921. A case of persistent hiccough. *Lancet, 197,* 171.

Sauter, D., Eisner, F., Ekman, P., and Scott, S. K. 2010. Cross-cultural recognition of basic emotions through nonverbal emotional vocalization. *Proceedings of the National Academy of Sciences, U.S., 107,* 2408–2412.

Schino, G., and Aureli, F. 1989. Do men yawn more than women? *Ethology and Sociobiology, 10,* 375–378.

Schmelz, M., Schmidt, R., Bickel, A., Handwerker, H. O., and Torebjork, H. E. 1997. Specific C-receptors for itch in human skin. *Journal of Neuroscience, 17,* 8003–8008.

Schmidt, W., Cseh, I., Hara, K., and Kubli, F. 1984. Maternal perception of fetal movements and real-time ultrasound findings. *Journal of Perinatal Medicine, 12,* 313–318.

Schoenwolf, G. C., Bleyl, S. B., Brauer, P. R., and Francis-West, P. H. (eds.). 2009. *Larsen's human embryology*. 4th. ed. Philadelphia: Churchill Livingstone.

Schurmann, M., Hesse, M. D., Stephan, K. E., Saarela, M., Zilles, K., Hari, R., and Fink, G. R. 2005. Yearning to yawn: The neural basis of contagious yawning. *Neuroimage, 24,* 1260–1264.

Scott, T. R., and Verhagen, J. V. 2000. Taste as a factor in the management of nutrition. *Nutrition, 16,* 874–885.

Seijo-Martinez, M., Varela-Freijanes, A., Grandes, J., and Vazquez, F. 2006. Sneeze area in the medulla: Localization of the human sneeze centre? *Journal of Neurological Surgery and Psychiatry, 77,* 559–561.

Selden, B. S. 1989. Adolescent epidemic hysteria presenting as a mass casualty, toxic exposure incident. *Annals of Emergency Medicine, 18,* 892–895.

Seligman, M. E. P. 1971. Phobias and preparedness. *Behavior Therapy, 2,* 307–320.

Senju, A. 2010. Developmental and comparative perspectives of contagious yawning. In O. Walusinski (ed.), *The mystery of yawning in physiology and disease,* 113–119. Basil: Karger.

Senju, A., Kikuchi, Y., Akchi, H., Hasegawa, T., Tojo, Y., and Hiroo, O. 2009. Does eye contact induce contagious yawning? *Journal of Autism and Developmental Disorders, 39,* 1598–1602.

Senju, A., Maeda, M., Kikuchi, Y., Hasegawa, T., Tojo, Y., and Osanai, H. 2007. Absence of contagious yawning in children with autism spectrum disorder. *Biology Letters, 3,* 706–708.

Seuntjens, W. 2004. On yawning, or the hidden sexuality of the human yawn. PhD dissertation, Vrije Universiteit, Amsterdam.

Sherrington, C. S. 1906. *The integrative action of the nervous system*. New Haven, CT: Yale University Press.

Shoup-Knox, M. L., Gallup, A. C., Gallup, G. G., and McNay, E. C. 2010. Yawning and stretching predict brain temperature changes in rats: Support for the thermoregulatory hypothesis. *Frontiers in Evolutionary Neuroscience, 2,* 1–5.

Simner, M. L. 1971. Newborn's response to the cry of another infant. *Developmental Psychology, 5,* 136–150.

Sloan, P. G. 1962. Ocular side effects of systematic medication. *American Journal of Optometry and Archives of American Academy of Optometry, 39,* 459–470.

Small, G. W., and Borus, J. F. 1983. Outbreak of illness in a school chorus. *New England Journal of Medicine, 308,* 632–635.

Small, G. W., and Nicholi, A. M. 1982. Mass hysteria among schoolchildren. *Archives of General Psychiatry, 39,* 721–724.

Smith, D. W. 1981. Mechanics in morphogenesis: Principles and response of particular tissues. *Recognizable patterns in human deformation,* Vol. 21: *Major problems in clinical pediatrics,* 110–144. New York: W. B. Saunders.

Smoski, M. J., and Bachorowski, J.-A. 2003. Antiphonal laughter between friends and strangers. *Cognition and Emotion, 17,* 327–340.

Smotherman, W. P., and Robinson, S. R. (eds.). 1988. *Behavior of the fetus.* Caldwell, NJ: Telford Press.

Soltis, J. 2004. The signal function of early infant crying. *Behavioral and Brain Sciences, 27,* 443–458.

Souadjian, J. V., and Cain, J. C. 1968. Intractable hiccup: Etiologic factors in 220 cases. *Postgraduate Medicine, 43,* 72–77.

Spinrad, T. L., and Stifter, C. A. 2006. Toddler's empathy-related responding to distress: Predictions from negative emotionality and maternal behavior in infancy. *Infancy, 10,* 97–121.

Spock, B. 1968. *Baby and child care,* rev. ed. New York: Pocket Books.

Sroufe, L. A., and E. Waters 1976. The ontogenesis of smiling and laughter: A perspective on the organization of development in infancy. *Psychological Review, 83,* 173–189.

Sroufe, L. A., and J. P. Wunsch 1972. The development of laughter in the first year of life. *Child Development, 43,* 1326–1344.

Stein, P. S. G. 1983. The vertebrate scratch reflex. *Symposium of the Society for Experimental Biology, 37,* 383–403.

Stevenson, R. E. 1999. Interview with Dr. Robert Stevenson. Johnson Space Flight Center Oral History 2 Transcript. JSC.nasa.gov/history/oral_histories/S-T.htm.

Straus, C., Vasilakos, K., Wilson, R. J. A., Oshima, T., Zeiter, M., Derenne, J.-P., Similowski, T., and Whitelaw, W. A. 2003. A phylogenetic hypothesis for the origin of hiccough. *BioEssays,* 25, 182–188.

Stromberg, B. V. 1975. Sneezing: Its physiology and management. *Ear, Nose and Throat Monthly, 54,* 449–453.

Sugiyama, L. S. 2004. Illness, injury, and disability among Shiwiar forager- horticulturalists: Implications of health-risk buffering for the evolution of human life history. *American Journal of Physical Anthropology, 123,* 371–389.

———. 2005. Physical attractiveness in adaptionist perspective. In D. M. Buss (ed.), *Evolutionary psychology handbook,* 292–343. New York: John Wiley.

Sullivan, D. A., Stern, M. E., Tsubota, K., Dart, D. A., Sullivan, R. M., and Bloomberg, B. B. (eds.). 2002. *Lacrimal gland, tear film, and dry eye syndromes 3.* New York: Springer.

Sun, Y.-G., and Chen, Z.-F. 2007. A gastrin-releasing peptide mediates the itch sensation in the spinal cord. *Nature, 448,* 700–703.

Sun, Y. G., Zhao, Z. Q., Meng, X. L., Yin, J., Liu, X. Y., and Chen, Z. F. 2009. Cellular basis of itch sensation. *Science, 325,* 1531–1534.

Symons, D. 1979. *The evolution of human sexuality.* Oxford: Oxford University Press.

——. 1995. Beauty is in the adaptations of the beholder. In P. R. Abramson and S. D. Pinkerson (eds.), *Sexual nature, sexual culture,* 80–118. Chicago: University of Chicago Press.

Thornhill, R., and Gangestad, S. W. 1999. Facial attractiveness. *Trends in Cognitive Science, 3,* 452–460.

Tomasello, M., Hare, B., Lehmann, H., and Call, J. 2007. Reliance on head versus eyes in the gaze following of great apes and human infants: The cooperative eye hypothesis. *Journal of Human Evolution, 52,* 314–320.

Treisman, M. 1977. Motion sickness: An evolutionary hypothesis. *Science, 197,* 493–495.

Turrigiano, G. 2004. A competitive game of synaptic tag. *Neuron, 44,* 903–904.

van Haeringen, N. J. 2001. The (neuro)anatomy of the lacrimal system and the biological aspects of crying. In A. J. J. M. Vingerhoets and R. R. Cornelius (eds.), *Adult Crying: A biopsychosocial approach,* 19–36. Philadelphia: Brunner-Routledge.

van Hoof, J. A. R. A. M. 1972. A comparative approach to the phylogeny of laughter and smiling. In R. A. Hinde (ed.), *Non-verbal communication,* 209–241. Cambridge: Cambridge University Press.

van Hoof, J. A. R. A. M., and Preuschoft, S. 2003. Laughter and smiling: The intertwining of nature and culture. In F. B. M. de Waal and P. L. Tyack (eds.), *Animal social complexity: Intelligence, culture, and individualized societies,* 261–287. Cambridge, MA: Harvard University Press.

Vettin, J., and Todt, D. 2004. Laughter in conversation: Features of occurrence and acoustic structure. *Journal of Nonverbal Behavior, 28,* 93–115.

——. 2005. Human laughter, social play, and play vocalization in non-human primates: An evolutionary approach. *Behaviour, 142,* 217–240.

Vincent, A. 2010. John Newsom-Davis. *ACNR (Advances in Clinical Neuroscience and Rehabilitation), 10,* 30–31.

Vingerhoets, A. J. J. M., and Cornelius, R. R. (eds.). 2001. *Adult crying: A biopsychosocial approach.* Philadelphia: Brunner-Routledge.

Vogel, D. H. 1979. Otolaryngologic presentation of tic-like disorders. *Laryngoscope, 89,* 1474–1477.

Vuorenkowski, V., Wasz-Hockert, O., Koivisto, E., and Lind, J. 1969. The effect of cry stimulus on the lactating breast of primipara: A thermographic study. *Experientia, 25,* 1286–1287.

Walshe, F. M. R. 1923. On certain tonic or postural reflexes in hemiplegia with special reference to the so-called "associative movements." *Brain, 46,* 1–37.

Walusinski, O. (ed.). 2010. *The mystery of yawning in physiology and disease*. Basel: Karger.

Watson, P. G., and Young, R. D. 2004. Scleral structure, organization and disease. A review. *Experimental Eye Research, 78,* 609–623.

Weisfeld, G. E. 1993. The adaptive value of humor and laughter. *Ethology and Sociobiology, 14,* 141–169.

Weiskrantz, L., Elliot, J., and Darlington, C. 1971. Preliminary observations on tickling oneself. *Nature, 230,* 598–599.

Wessel, M. A., Cobb, J. C., Jackson, E. B., Harris, G. S., and Detwiler, A. C. 1954. Paroxysmal fussing in infancy, sometimes called "colic." *Pediatrics, 14,* 421–434.

Whitman, B. W., and Packer, R. J. 1993. The photic sneeze reflex. *Neurology, 43,* 868–871.

Wilson, B., Batty, R. S., and Dill, L. J. 2003. Pacific and Atlantic herring produce burst pulse sounds. *Biology Letters, 271,* S95–S97.

Winkworth, A. L., Davis, P. J., Adams, R. D., and Ellis, E. 1995. Breathing patterns during spontaneous speech. *Journal of Speech and Hearing Research, 38,* 124–144.

Wolff, P. H. 1969. The natural history of crying and other vocalizations in early infancy. In B. M. Foss (ed.), *Determinants of infant behavior,* 81–108. London: Methuen.

——. 1987. *The development of behavioral states in the expression of emotions in early infancy: New proposals for investigation*. Chicago: University of Chicago Press.

Wong, R. O. 1999. Retinal waves and visual system development. *Annual Review of Neuroscience, 22,* 29–47.

Wynne-Jones, G. 1975. Flatus retention is the major factor in diverticular disease. *Lancet, 306,* 211–212.

Yanoff, M. 1969. Pigment spots of the sclera. *Archives of Ophthalmology, 81,* 151–154.

Yoon, J. M. D., and Tennie, C. 2010. Contagious yawning: A reflection of empathy, mimicry, or contagion? *Animal Behaviour, 79,* e1–e3.

Zahn-Waxler, C., Friedman, S. L., and Cummings, E. M. 1983. Children's emotions and behaviors in response to infants' cries. *Child Development, 54,* 1522–1528.

Zilli, I., Giganti, F., and Uga, V. 2008. Yawning and subjective sleepiness in the elderly. *Journal of Sleep Research, 17,* 303–308.

acknowledgments

I thank a generation of mostly undergraduate research students who contributed to this work, especially Yvonne Yong, Kenneth Fisher, Heidi Hamernik, Barbara Curchack, Lisa Greisman, Nicole Brocato, Robert Spencer, Kurt Krosnowski, Darcy Mandell, Marcello Cabrera, Jessica Nave-Blodgett, Skylar Spangler, Katie Webb, Clifford Workman, Tina Runyan, Megan Hosey, Kimberly Provine Lourenco, Susan Conn, Bentley Tate, Jason Tate, Lisa Geldmacher, Bernard Fischer, and David Nodonly. Judy Westerman, Kit Strawbridge, Barbara Harrison, and Deborah Tracy provided dedicated service before this specific line of work got underway. Many are now doctors and professors of various kinds, both PhD and MD. All are exceptional in their own ways.

Paul Rozin, an inspiration for a generation of physiological psychologists, shared his wisdom about things disgusting—most people find these matters, well, disgusting. Terence Anthoney shared his unique knowledge of hiccupping. The book is better because of comments on one or more chapters by

Paul Bloom, Paul Rozin, James Kalat, and Ron Oppenheim. Hanneke de Vries contributed valuable insights about the prenatal origin of several behaviors. The wisdom of Jaak Panksepp, Gordon Gallup, Ron Barr, Brian Hopkins, William Frey, and Roger Fouts was always helpful. Steve Pinker encouraged my shrinking of the blank slate. Kim Bard generously shared her expertise and provided access to her young chimpanzee charges when she was employed at the Yerkes Primate Center in Atlanta. UMBC colleague Bernie Lohr provided access to his lab and assisted in acoustic analysis; Gene Morton provided a similar service two decades earlier at the National Zoo. Tim Ford provided expert graphic assistance for many projects. Tom Dunne contributed most of the artwork in this book, some of which is based on articles by the author originally published in *American Scientist* magazine.

The influence of Rita Levi-Montalcini and Viktor Hamburger, two of my mentors, is found throughout, especially in Chapter 13, Prenatal Behavior. Both are models for a long and productive life in science; Viktor reached his 100th birthday and Rita is now 102, the oldest Nobel Laureate. Viktor reminded his students that the foremost criterion for the Doctor of Philosophy degree is scholarship; I hope that this book would have passed his stern test, but he may have been challenged by my occasional whimsy. Thomas Sandel encouraged me to pursue my interests, wherever they led; Jerry Lettvin encouraged only research that would change everything, a daunting standard.

My greatest thanks go to my wife, pianist Helen Weems, whose keen ear extends to prose. Helen read the entire manuscript and provided invaluable editorial assistance at every stage of writing. Her stoic acceptance of the austere role of a writer's wife is remarkable and appreciated. This book is dedicated to her with love and gratitude.

index